AF531681

ENCYCLOPAEDIA OF PLANT PROTECTION

ENCYCLOPAEDIA OF PLANT PROTECTION

Vol. 1

B.A. Mannan

ANMOL PUBLICATIONS PVT. LTD.
NEW DELHI - 110 002 (INDIA)

ANMOL PUBLICATIONS PVT. LTD.
H.O.: 4374/4B, Ansari Road, Darya Ganj,
New Delhi-110 002 (India)
Ph.: 23278000, 23261597
B.O.: No. 1015, Ist Main Road, BSK IIIrd Stage
IIIrd Phase, IIIrd Block,
Bangalore - 560 085 (India)
Visit us at: www.anmolpublications.com

Encyclopaedia of Plant Protection

First Published, 2008

ISBN 978-81-261-3538-7 (Set)

PRINTED IN INDIA

Printed at Mehra Offset Press, Delhi.

Contents

PREFACE

Plant lovers, food producers and agricultural planners, all voice together to protect the plants from insects and diseases. Damages to the plants by insects are easily felt, but not realised fully. Every year, hundreds of people die of hunger. We can provide them with enough, food, if food crops are kept safe from the damages.

Plants, like human beings, are prone to different diseases. A diseased plant cannot yield good produces. Hence, plant protection has attained greater significance in today's world. Agricultural crops have been prone to damage, because of insects, rodents, diseases, weeds and nematodes, etc. since time immemorial.

Plant Protection continues to play a significant role in achieving targets of crop production. Major thrust areas of plant protection are promotion of integrated pest management, ensuring availability of safe and quality pesticides for sustaining crop production, despite the ravages of pests and diseases by streamlining the quarantine measures for accelerating the introduction of new high-yielding crop varieties, besides eliminating the chances of entry of exotic pests.

It is essential to have the basic knowledge about the plant pathogens, different kinds of diseases, caused by them to various crops and of course, the methods of controlling diseases to avoid enormous losses caused by the diseases. As plant protection is indispensable, job opportunities are always there for people, having basic knowledge of pests, crops and diseases.

This book is a serious endeavour in that direction. This *Encyclopaedia of Plant Protection,* in four volumes takes the stock of present state of art in plant protection. It also focuses on isolation, structure, characterisation and physiochemical and physiological properties of plants, synthesis, functions and mechanisms of action of plant protease inhibitors. It also deals with their inhibitory effects on insects and pathogens and with the resultant application of genetic engineering to plants for their protection.

This book caters to the need of knowledge hungry biologists, agriculturists, policy planners and specialists in the field of plant protection and of course the students, teachers, researchers and beyond that, all those concerned with agricultural production and its development.

Editor

1

PESTS' ROLE

It is estimated that insects evolved on earth about 250-500 million years ago whereas man has appeared only 1 million year back. This gives an idea about the immense degree of evolutionary maturity of insects in comparison to man. The number of species of insects has also always been greater than the total number of species of all other forms of life put together. As per literature, out of 9 lakh animal species more than 6 lakh belong to class Insecta which roughly constitute more than 70 per cent of the total forms of living creatures. This analysis shows that insects have been fully exploiting their characteristic minuteness of their best advantage in the struggle for existence and adapted them for the extreme environment. Till very recently the public was hardly concerned about the crop pest's infestation except in cases of very severe epidemic and that also used to forget as soon as the epidemic subsided.

In fact the realty is that "no other animal sustain such intimate and complex relations to plants as insects do. "The more luxuriant and varied the flora, the more abundant and diversified its accompanying insect fauna" — Felson. It is well known that insects affect human interest both for good and evil, but their destructive nature is better known to us than their beneficial group of insects, there would have a lot of shortcomings and problems in our society because we get many things of daily use from them, *viz.*, cloths (silk), honey, lac, wax and medicines etc. Besides, some of the insects like queen termite, locust and larvae of many insects are used as food in different parts of the world.

Considering the role of insects in Agriculture, they can be mainly classified into three groups *viz.*, harmless, injurious and beneficial insects. The first group includes hundreds of insect species may be over looked as it dose not affect the economy of man. The remaining two groups may be conveniently arranged into different categories according to their destructive or useful nature. The following is the flow chart which gives picture of harmful and beneficial insects that directly concern the role of insects in agriculture:

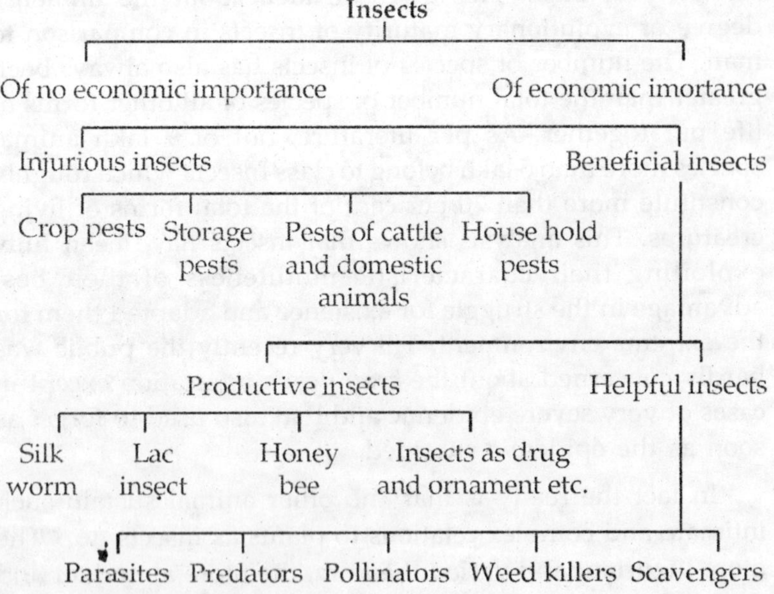

Harms to Agriculture by Insects

The damage caused by the insects to agriculture may broadly be grouped into two *viz.*, direct and indirect losses:

Direct Losses to the Plants

Root: The insects which live in soil (soil inhibitors) cause considerable damage to the roots of plants, *e.g.*, termites, beetles and their grubs, ants and larvae. Some of them completely destroy the roots resulting in the death of plants, while some species eat only the portion of the roots due to which plants becomes weak and sickly in appearance.

Stem: Many caterpillars enter the stem of various plants and cause very serious damage (stem borers), while others cut the growing part and check the growth of plants, *e.g.*, grasshoppers, termite, cricket and shoot flies etc.

Leaf: Leaf eating insects may cause the damage by various way as described below:

(i) *Leaf Eaters or Defoliators:* The insects having biting and chewing type of mouth parts and feed on the leaves may be put under this category, *e.g.*, grasshoppers, locusts beetles, weevils and caterpillars of moths and butterflies. Some time the damage is so severe that they completely defoliate the crops leaving only the bare stems.

(ii) *Leaf Miners:* They live between the upper and lower epidermis of leaves and devouring it's green part. The mining habit is generally adopted by larvae of some insects, *e.g.*, pea leaf miner *(Phytomyza atricornis)*, citrus leaf miner *(Phyllocnistris citrella)*

(iii) *Leaf Rollers:* Some of the larvae besides feeding upon the leaves cause them to roll up which often some times shrivel and fall, *e.g.*, cotton leaf roller *(Sylepta derogata); juar* leaf rollar *(Marasmia trapezallis)*.

(iv) *Sap Suckers:* Many insect species having piercing and sucking type of mouth parts drain the sap from leaves due to which leaves become brown or yellow and fall, *e.g.*, aphids, sugarcane white fly, pyrilla, 'mango leaf hopper etc.

Flower and Fruit Feeders: Among the flower feeders blister, beetles, thrips and jassids are the important pest whereas the fruits are damaged by a variety of insect pests namely fruit sucking moths, fruit fly and fruit borers. The damage caused by these pests lowers the quality as well as the quantity of fruits.

Seed Feeders and Insects of Godowns: There are some insects which feed either on the developing seeds or the mature seeds in plants. The striking examples of such insects are tobacco caterpillar and gram pod borer in sunflower. However, certain insects *viz.*, rice weevil and pulse beetle, which cause damage to seeds in field as well as in godowns. The insects that damage to seeds only in godowns are rust red flour beetle, khapra beetle, lesser grain borer, rice moth and saw toothed grain beetle etc.

Wood and Bark Feeders: Some species of insects feed the bark and wood of the plant, as a result of this the quality of wood which is used in buildings and furniture etc. is reduced appreciably. The main insects belonging to this group are the beetles, weevils, bark eating caterpillar and termites etc.

Indirect Losses to the Plants

Besides causing direct loss to the plants there exists some insect species which produce honey dew from their body while draining the sap from plant system and this results in the formation of moulds. These moulds interrupt in the photosynthetic and respiratory activity of the plant and thus finally the growth is adversely affected. The example of such insects are hoppers and aphids.

In addition to this, a number of insect species are known which transmit viral and mycoplasma like diseases in plants. The insect vectors which belong to this group are sesamum, jassids, white flies, aphids and grasshoppers. Leaving aside the indirect losses to the plants, insects also transmit the disease like malaria, typhoid, cholera, elephantitis and dysentery to human being. Some of them also carry over the pathogens which cause diseases to the livestock, *e.g.*, the bite of tse-tse fly causes Nagana in the cattle and sleeping sickness in the human being.

Thus, it is evident that every year a large portion of agriculture produce is damaged by the insects. Roughly it has been estimated that in India annual loss caused by the insects is to the tune of Rs. 500 million, out of which 50 million to Uttar Pradesh alone.

Utility of Insects for Agriculture

As stated earlier that all insect species are not the enemies of agriculture but some of them are allies of man, which has been discussed as under.

Direct Advantages

There are some insects which are of great economic importance as they produce valuable things like honey, silk, lac, ornamentals and medicines to the man. The annual income from these articles in India is about 50 crores of rupees. These insects are called as beneficial insects and have been dealt in separate chapter.

Indirect Advantages

There are indirect advantages also from the insects which we feel in our daily life and that are given below:

Pollination: It is well known fact that most of the plants are cross pollinated and it has been estimated that 5% flowers are self pollinated, 10% by wind and water and 85% are pollinated by insects which clearly indicates the value of insects and without them many plants do not thrive well, *e.g.*, Anjeer, Gular, Pipal and Bargad. Several other plants of economic importance also depend upon insects to help in pollinating the blossom for a good yield. The flowers of cherries, pear and peach depend on the insects for their reproduction.

Parasites and Predators: The insects which lay their eggs on or within the body of insect's host, and their immature stages feed on or within the body of the hosts are known as parasites, where as those which catch larvae or other living organisms and devour them are called predators. The insects are always present in the nature and playing a vital role in keeping the insects pest population low and when the population of parasites and predators is reduced due to certain reason, an insect pest may appear in epidemic form. The role played by these insects has been discussed in the chapter of biological control. Some of the outstanding examples are given below:

Mantis	predator
Coccinellids	do
Lady bird beetle	do
Trichogramma spp.	Parasite
Apantalis	do

Drugs and Ornaments: Some specific drugs are prepared from the insects, *e.g.*, powdered dry cockroaches are said to be useful in urine trouble, blister beetles are good source of cantharidine which are fed to the cattles for removing sterility. Ancipient paralysis is caused by the anaphelese mosquito infected with the malarial parasite. Allantpini chemical is recovered from the larvae of housefly which has got the healing power of wounds. There are many insects which are used as ornaments, *e.g.*, glow worms are enclosed in gauze and used as hair ornaments.

Scavengers: There are certain insects, which act as scavengers by way of eating waste materials, *e.g.*, flies. The dung beetle is another example which roll some portions of the dung etc. and burry in the ground. This serves the purpose of fertilizing the soil. Some insects break up the dead wood and act as a scavengers of dead animal matter.

Weed Killers: The weeds which are harmful to crop, are kept under control by insects by way of feeding on them, *e.g.*,

Lanlana camera is controlled by a moth, *Playtilia pussidactyla.* Cactus is controlled by moth, Cactoblastis cactorum.

Aṣ a Food: Some insects are used in the form of food, *e.g.*, queen termites, locusts and larvae in different parts of the world.

It is clear from the above that insects besides causing damage, have certain advantages also from them and in this respect they are useful to the farmers.

Non-insect Pests of Agriculture

There is a group of animals, other then insects, which cause considerable loss to the agricultural crops and commonly called as non-insect pests. It includes mammals, rodents, birds, molluscs, mites and nematodes etc.

Mammals

Damage done by various mammals to agriculture has not been estimated properly, but there is no doubt that they cause considerable loss to agricultural and horticultural crops during various stages of their growth. The following animals of this group are more important.

Monkey: The three species namely; *Rhelus machaque; Macaca mulatto* and *Semmopithecus futellus* are most common in India. According to an estimate there are about 60 lakh monkeys in India out of which 1/3 are found in U.P. alone. They do more damage than actually what they eat by way of chewing and biting the fruits etc. The favourite foods of the monkeys include; sugarcane, maize, sorghum, gram, peas, beans, mango, guava, Jamun, banana, papaya and groundnut etc.

Control: They can not be killed due to religious sentiments, therefore, may be caught in traps and sent to the other places. Dogs chase them away, hence may be encouraged for this purpose.

Blue Bulls (Boseelaphus tragocamelus): *The* blue bull commonly known as 'Neelgi' is found in U.P., Bihar, M.P., Punjab, Mysore, Mumbai, Jammu-Kashmir and Delhi and mostly live in herd. They are a great menace to agricultural crops with particular reference to sorghum, bajra, maize, paddy, wheat and gram. They feed, graze and cause lot of damage by trampling. They generally feed at night.

Control: Fencing around the field is most effective method to control them.

Deer: The four species *viz.; Antelope cervicarpa; Gazela hennetti; Axis axis* and *Cervus unicolor* are found in India. They generally live in herd and graze various crops. Besides feeding a great deal of damage is done with their hoofs in compacting soil. They also cause damage to plants by rubbing their horns on stem of young trees stripping off the bark and often breaking the trees.

Control: Watching and fencing around the field.

Jackal (Cants aureus): They are nocturnal in habit and found all over the country except Himachal Pradesh. Jackals live in herds and mainly cause damage to maize, sugarcane and groundnut fields. Besides, they do some damage to poultry by way of eating the chickens.

Control: Poison baiting, fencing are the practical methods of their control. About an ounce of zinc phosphide or strychnine hydrochloride may be placed on the small pieces of meat and kept in places normally visited by them.

Wild Boar (Susscrofa cirstatus): It is generally found in Assam, Bihar, U.P., Rajasthan, Maharashtara, A.P., Kerala and West Bengal. They feed and uproot the crops by their snout and cause considerable loss. The tuber crops *viz.* ; potato, sweetpotato and groundnut are specially damaged. The bear *(Melursus urusinus)* attacks sugarcane and maize crops and climb the palm tree to drink toddy from the pot hung for collecting it. They also relish to drink honey from the comb of bees.

Control: Shooting is the only practical method to control them. The crops may also be protected by fire works.

Porcupine (Hystrix indica): It is found in all the states except Andhra Pradesh, Bihar, Tamil Nadu and Karnataka.

They usually live in burrows and are nocturnal in habit. The furs are modified into spines and hollow quills. The porcupines cause damage to all kinds of vegetable with particular reference to root crops (potato, sweetpotato and groundnut).

Control: Poison baiting with zinc phosphide in guava slices has been found effective. Fumigation of burrows with cyno-gas may also be practiced.

Rabbit and Hares (Lepus nigricollis* and *ruficaudatus): Rabbits and hares are found in all the states except Bihar, Tamil Nadu, West Bengal and Manipur. They live in burrows nocturnal in habit and cause damage to crops in the seedling stage.

Control: Fumigate the burrows with cyano gas.

Squirrel: The two species namely ; *Funambulus pennati and F. plamarum* are generally found in India. They are diurnal in habit and feed on the flowers, fruits and nuts.

Control: Poison baiting with zinc phoshide or shooting.

Flying Foxes: The two species, *viz., Pteropus giganteu sand, Cynopterus sphinx* are more common in India. They are found all over the country except West Bengal and Manipur, Flying foxes are nocturnal in habit, vegetarian and feed gregariously. They mainly feed on fruits (guava, mango, litchi, jamun, banana, ber and black berry) thus are the great enemy of the gardens. The damage caused by them is estimated to the tune of 12 to 40 per cent.

Control: They may be caught by hanging the net in the orchards. They may be kept away by hanging the electric bulbs on the top of the tree or by making noise.

Birds

The birds out number all other vertebrates except fishes and there are about 8,600 species of birds found in different parts of world (Salim Ali, 1972). Some species of birds are harmful to agriculture but actual damage caused by them is not exactly known. They cause damage to growing crops, fruit trees in the fields, orchards, threshing yards and in houses. Some harmful birds are described here as under:

Crow: Two species of it namely, *Corvus splendens* and C. *macrorhynchos* are commonly found through out the country. They have been reported to cause heavy damage to maturing and ripe crops of agricultural and horticultural importance with particular reference to groundnut, sorghum, chillies, wheat, fig, mango and guava. They are menace to poultry farming as they take away young chicken and eggs and to the live stock breeder as they peck the eyes out of new born lambs.

House Sparrow (Passar domesticus): It usually lives and nests in the residential building. The adult feeds on the crumbs of man's dining table but the young-ones are fed on insect larvae. When their population is high, they cause severe damage to crops like wheat, sorghum, millets and small-sized succulent fruits.

Parrot: Many species of it are found in India, out of which 3 species *viz.*; *Psittacula krameri; Peupatria nepalensis* and *alexandri fasciatus* are more common. It is very destructive bird that wastes more than it eats. They eat and cut the ripe and semi-ripe fruits and cause heavy damage to agricultural and horticultural crops with particular reference to maize, sorghum, wheat, gram, mango, jamun, guava and other fruits.

Other Birds: There are some other birds also which cause damage to various agricultural and horticultural crops. The important one are—Bulbul *(Pycnontus cafer)*; Baya *(Ploceus philippinus)*; Myana *(Acrudotheres tristis)* ; Pigeon *(Coluba livia)* and green bee-eater *(Merops orientalis)*. These birds mostly feed on grain crops, newly sproution vegetables and fruits of various kinds.

Control: Watching and producing sound are the best methods to control the birds. They can be caught with net-traps. In some cases poison baiting and repellent chemicals may be effective.

Molluscs

Mollusca is the second largest phylum of animal kingdom, the largest being Arthropoda. It includes snails and slugs which are important from agriculture point of view. The important characteristic of the phylum Mollusca are given below:

(i) They are unsegmented and without appendages.

(ii) They are bilaterally symmetrical and have three germ layers.

(iii) Body is covered by a thin dorsal mantle which secretes the limy shell.

(iv) Mostly aquatic in habit but some of them are land dwellers.

(v) Respiration usually takes place through gills.

(vi) Circulatory system is open.

(vii) They are usually unisexual but some like snails are bisexual.

(viii) Excretion is nephridial.

(ix) The development may be direct larval.

(x) Snail possesses a shell which is made up of one piece while slug has no shell.

Snail

The following species of snail are commonly found:

Viriparus vairatus, Pila virens

Achantia fulica, Indoplanorbis exuslus

Achantia fulica, commonly known as Giant African Snail, is a serious pest of agricultural and horticultural crops.

It is found in different parts of the world including India and islands of the Andaman and Nicobar group. They are nocturnal in habit and during day time they hide under debris and bushes. They are the pest of paddy, vegetables fruits, ornamentals and plantation crops like banana and papaya etc. Paddy crop is seriously damaged by them in Kashmir and some parts of U.P. and South India. They feed on the leaves and stems of the plant.

Control: Bait poisoning with 6% lead arsenate or 5% metaldehyde or 3.8% Paris green in wheat bran and mollases have been found effective. The two species of snails are the predator of giant African snail and they may be utilized in control programme:

(i) Eugladina rosea,

(ii) Gonaxis quadrilateralis.

Slug

The slugs feed on either dead or living vegetation and live in the moist damp places. Like snails, they are nocturnal in habit. They have been observed to damage vegetable crops and banana etc. in U.P., Maharashtra, Assam and M.P.

They cut the growing plants near the root surface with the results plants die. In Mysore the slugs have been found to damage young citrus plants and some times they completely defoliate the citrus seedlings.

Control: The wood ash may be broadcasted near the roots of plants to protect the damage. Bait poisoning has also been found effective. The use of kerosine-oil has been reported to give the control of slugs.

Mites

Mites belong to the phylum Arthropoda, class Arachnida and order Acarina. The following are the main charaterstics of this group:

(i) They are small creatures measuring 0.1 mm long.

(ii) The body is divided into two:

(a) prosoma-head and thorax fused together,

(b) opisthosoma-abdomen.

(iii) Mouth parts are chewing, piercing and sucking type.

(iv) Legs are two segmented with chelate or padipalpi.

(v) Larval and nymphal stages possess 3 pairs of legs while adults have four pairs of legs.

(vi) Respiration takes place by trachesae which are connected with stigmata. In some cases it may be cutaneous.

(vii) Metamorphosis is simple.

(viii) The male is smaller than female.

Economic Importance and Nature of Damage

Mites are found in almost every habitat. They not only damage the plants but are also parasite on human beings and domesticated animals and responsible for various diseases. Phytophagous mites feed by inserting stylet like chelicerae into the host cells and suck the plant juice.

They mostly feed on the leaves with the result leaves become yellow in appearance and fall down after sometime. In mango they suck the juice from the inflorescence, therefore, fruit bearing is adversely affected. Mites are also carriers of fungal and viral diseases of plants.

Life History

Usually all the species of mites have similar type of life cycle, only there is a difference in the time of attack which depends upon the season and time of crops. The female lays round eggs after copulation on the lower surface of the leaves singly or in clusters. Hatching period of eggs varies from 3 to 7 days. After hatching tiny larva comes out which has 2 or 3 pairs of legs. The larva before changing into adult has two nymphal stages namely parrot and due to nymphs. They generally similar to adult in size and colour. The larval period lasts 10-15 days. The adult mite has four pairs of legs except Eriophyes mite which possesses only two pairs of legs. Their life cycle is completed in two to three weeks. Under favourable conditions they grow very fast in number and spread in all over the field. When the season is unfavourable they live either in egg or adult stage. The extensive use of chemical and fertilizer is probably responsible for their out breaks.

The following are the important species of phytophagous mites commonly founds in India:

1. Sugarcane mite *(Oligonychus indicus, Schizotetranychus adnropogoni).*

2. Red spider mite *(Tetranychus bioculatus)*
3. Vegetable mite *(T. cucurbitae)*
4. Okra mite *(T. telarius)*
5. Mango mite *(Aceria mangiferae)*
6. Citrus mite *(Eutetranychus banks!)*
7. Litchi mite *(Eriophyes litchi) Non Insect Pests of Agriculture (365)*
8. Castor mite *(Anichus ricini)*
9. Cotton mite *(Eriophyes gossypii)*
10. Sorghum mite *(Eutetrahynchus indicus)*
11. Paddy mite *(Paratetranychus oryzae)*
12. Wheat mite *(Petrobia latens)*
13. Apple mite *(Cenopaopus pulcher)*

There are also some mites which damage the stored grains during storage and some of them are as follows:

(i) *Acarus siro*
(ii) *Tyrophagous putrescentiae*
(iii) *Acaropsis docta*
(iv) *Acaroyhenax trieboli*
(v) *Blattisocius tarsalis*

Control: The mites are generally controlled by the application of certain pesticides, and some of them are specific and called acaricides or miticides. The commonly used acaricides are given below:

Sulphur: Sulphur is one of the earliest chemicals used for the control of mites. It is specially used against false spider mite responsible for leaf curl disease of litchi. It is used in the form of dust, wettable powder or sulphur solution.

Zineb or Dithane: It is primarily a fungicide but has been found effective to control the *Tetranychid* mites attacking various fruit trees. In America it is found very effective to control the citrus mite.

Aramite: It is chemically 2-(P-tertra butyl phenoxy) phenoxy 1-2 chloroethyl sulphite. The technical material is brownish oil, boiling point 175° C and contains atleast 90% of the active ingredient. It is insoluble in waver but soluble in most organic solvents. Under alkaline conditions it hydrolyses and under strong sunlight liberates SO_2 Aramite is formulated as 15% Wettable powder, 25% EC and 5% dust. It is one of the most broadly effective acaricides.

Kelthane: Chemically it is 1-1 bis(P-chloropheny) 2, 2, 2-Trichloroethane. It is white crystalline solid, m.p 78.5-79.5°C. It is insoluble in water but soluble in organic solvents. It is formulated as 25% wettable powder and emulsifiable concentrate. Kelthane is found to be toxic to most species of *Tetranychid* mites, many species of Eriophyid mites and is highly effective against false spider mites.

Tedion: Chemically it is 2, 4, 5, 4-tetra chlorodiphenyl-sulphone. It is crystalline solid m.p. 148° C insoluble in water but soluble in common organic solvents, It is very stable to the action of alkalies and acids light and temperature has a very prolonged residual action. It is formulatd as 25% wettable powder and emulsifiable concentrate. It is effective against Teatranychid mites, Eriophyid and false spider mites.

Ovex or Ovotran: It is chemically P-chloro-phenyl-p-chloro benzene sulpholate. Ovex is a white solid, m.p. 86.5°C insoluble in water, slightly in alcohol but readily soluble in common organic, solvents. It has been extensively used in different parts of the world for the control of Tatranychid mites on a variety of ornamental plants and cultivated crops. It is most effective against newly hatched larvae. Ovex is formulated as 50% wettable powder.

Other Acaricides: There are also some other acaricides, *viz.*, sulphonone, chlorobenzide, flour-benzide, azobenzene, neotran DMC and chlorobenzlate which are used to control the different species of phytophagous mites.

Some insecticides namely phosphamidon, thiometon, parathion, malathion, phorate, diazinon and carbophenothion, have also found to be effective in controlling the mites.

Plant Parasitic Nematodes

Plant parasitic nematodes are generally transparent, unsegmented, bilaterally symmetrical, pseudocoelomate and triploblastic worms measuring from 0.5 mm to 18 mm in length. The adult males are always slender worms while females of some species have greatly rounded bodies, sometimes nearly spherical, but always with distinct neck.

The following are the main characteristics of these worms:

(i) They are unsegmented

(ii) Body is bilaterally symmetrical and triploblastic.

(iii) There is no true body cavity (pseudocoelomate).

(iv) Alimentary canal is in the form of straight tube ; the anterior portion opens in the mouth while posterior in the anus.

(v) Circulatory and respiratory systems are absent.

(vi) Excretory and nervous systems are well developed.

(vii) Locomotory organs, cilia, are absent.

(viii) There is no metamorphosis.

Habit

Plant parasitic nematodes take their food only from living plants and are obligate parasite. They can be distinguished from non-parasitic forms by the presence of a stylet or spear which is quite variable in shape. They are found in almost every kind of environment but cannot be active unless there is sufficient free water in their surrounding. The nematodes penetrate their stylet by repeated thrusts in the plant tissues and inject their saliva therein from the oesophageal glands. Then they suck the juice, perhaps partially digested from the host by intermittent pulsations of the oesophageal bulb or by contractions of the whole oesophagous.

Nature of Damage

As mentioned above the plant parasitic nematodes are characterised by their mouth spear with which they pierce the plant tissues in the same way as the sucking insects penetrate plants with their mouth parts. Symptoms of nematode infested plants are mostly non-specific and the superficial examination of the crops seldom permits the diagnosis of nematodes as the causative agent of the disease. The general symptoms are poor growth, stunting patchiness, discoloured foliage (yellowing) and wilting in the warmer part of the days which may also be produced by other organisms or by nutritional deficiencies. Some nematode species produce specific symptoms on different parts of plants such as gall formation on roots by root-knot nematode, cockles in place of grain by seed nematode ; lesions on the leaves by leaf and bud nematode. Such types of symptoms depend upon the nematode species and type of plant.

Life History

The life history of plant parasitic nematode is usually simple and similar in all the cases. The female nematode lays eggs either on the plant tissues or soil or on the gelatinous-mass outside the body. The egg undergoes a series of cell division forming the first stage larva. The first moult takes place inside the egg and the second stage larva emerges from the egg. The second stage larva is generally similar to adult except that it is small in size and devoid of developed reproductive organs. It usually remains in the same stage until it finds a suitable host. After finding a host, it starts feeding and passes through their additional moults. Thus there are four larval stages and after last moult, the nematode becomes the adult. The adult female feeds on the host plant and after copulation lays eggs.

Some Important Species of Plant Parasitic Nematodes

The plant parasitic nematodes may be broadly classified into two groups:

Above Ground Feeders: The category includes following nematodes:

(i) Seed gall nematode *(Anguina tritici)*

(ii) Leaf and bud nematode *(Aphelenchoides spp.)*

(iii) Stem and bulb nematode *(Ditylenchus spp.)*

Below Ground Feeders: They may be again divided into three groups as endoparasite, semi-endoparasite and ecto-parasite on the basis of mode of feeding.

Endoparasites

They may be sedentary or migratory *viz.*

Sedentary:

(i) Root-knot nematode (*Meloidogyne* spp.)

(ii) Cyst forming nematode (*Heterodera* and *Globodera*)

Migratory:

(i) Lesion nematode (*Pratylenchus* spp.)

(ii) Burrowing nematode (*Radopholus similis*)

(iii) Rice-root nematode (*Hirschemanniella* spp.)

Semi-endoparasites

Sedentary:

(i) Citrus nematode (*Tylenchulus semipenetrans*)

(ii) Reniform nematode (*Rotylenchulus reniformis*)

Migratory:

(i) Spiral nematode (*Helicotylenchus, Rotylenchus*)

(ii) Lance nematode (*Hoplolaimus* spp.)

(iii) Stunt nematode (*Tylenchorhynchus, Merlinius*)

Ecto-parasites

Sedentary:

(i) Ring nematode (*Criconema, Criconemoides*)

(ii) Sheath nematode (*Hemicriconemoides, Hemicycliophora*)

(iii) Pin nematode (*Paratylenchus* spp.)

Migratory:

(i) Needle nematode (*Longidorus* spp.)

(ii) Dagger nematode (*Xiphinema* spp.)

(iii) Stubby nematode (*Trichodorus* spp.)

(iv) Awl nematode *(Dolichodorus* spp.) These nematode species also transmit various viral diseases.

Wheat Gall Nematode (Anguina Tritici)

This was the first plant parasitic nematode recorded by **Needham** in 1743 and found through out the wheat growing parts of the world. It causes ear-cockle or mania disease. Besides wheat, the nematode produces gall in rye, spelt and emer. The nematode is also responsible for causing yellow-ear rot (tundu) disease of wheat in association with bacterium, *Clavibactor (Corynebacterium) tritici.*

Symptoms

The following are the important symptoms produced by the nematode:

1. The seedlings are swollen at the base and growing tips are twisted.
2. Affected plants are more or less stunted.
3. The leaves are wrinkled, rolled, twisted and curled.
4. The glumes of affected ear are spread and remain green for longer time.
5. Diseased heads are usually shorter and thicker.
6. Some heads contain only gall while others contain both galls and kernels.
7. In the case of tundu disease, the whole ear is covered with yellow mass of bacterium and there is no gall or kernel formation at all.

Life Cycle

Under natural conditions dry galls containing thousands of second stage larvae fall to the ground from ripe seed heads together with healthy seeds. In general farm practice the gall usually are sown with the seed. The gall becomes soft on absorbing moisture and the larvae are set free into the soil. From there, they reach the growing plants. The larvae are located between the young leaves that enclose the growing point and feed as ecto-parasite. Later on, as the plants grow and reach earring stage, the larvae penetrate into the primordia of the flower buds and stimulate the development of galls instead of normal seeds.

The nematodes in developing galls quickly grow to maturity into males and females. A single gall may have 40 females and same number of males at this stage. The female after mating with male inside the gall starts egg-laying and lays large number of eggs The first stage larvae develop inside the eggs and hatch in the form of second stage. These second stage larvae remain in dormant stage inside the galls. There is only one generation found in a year.

Control

1. The galls may be separated through the sieves.
2. Since the galls are lighter than wheat seed they may be separated by floating in 2% brine solution. The seed should be washed with fresh water and dried before sowing.

Cyst Nematode

The following species are of cyst forming nematodes.

(a) *Globodera*

(b) *Heterodera*

(c) *Meloidodera*

(d) *Cactodera*

(e) *Sarisodera*

Among these, *Globodera rostochiensis and Heterodera avenae* are more important economically. The former species is restricted to Nilgiri hills only, while later species is common in Rajasthan, Delhi, Haryana and Punjab, U.P. states. Therefore, here the description of *Heterodera avenae* is given.

Cereal Cyst Nematode (Heterodera Avenae)

This nematode species causes molya disease of wheat and barley. It may also attack oat and rye. The damage is more pronounced in barley than wheat.

***Symptoms*:** The important symptoms of the disease are given below:

1. Seedling plants may show injury as the fourth leaf appears with a slightly red tip.
2. Roots are short with multiple branching giving a bunchy appearance and often bearing small gall like formations.
3. Plants appear stunted, pale yellowish green in colour and never grow beyond 30 to 60 cm in height.
4. The tillering is reduced, ear heads and ears are reduced and finally no grain formation.
5. The plants that escape early damage produce short stalks and ears and yielding very poor.

Life History

The nematode passes unfavourable season in the form of cysts generally in the soil. A cyst consists of the dead body of female cuticle protecting the eggs which are formed within the body. Under favourable conditions the eggs hatch within the cyst and the larvae are set free into the soil in the form of second stage which is the infective stage. They enter inside the plant roots mostly from the tips of young roots. Invading larvae take up a position parallel to the central cylinder in small roots, usually with their head pointing towards the plant. In larger roots they remain near the cortex. The larvae feed and moult, become some what cylindroid and similarly reach to the IVth stage larvae. At this point in the life-history the sexes can not be differentiated, but as the development proceeds the female becomes flask shaped, while male remains more or less cylindroid. The male after assuming the final cylindrical shape moults for the fourth time and becomes an adult.

The female increases in girth and changes into more or less oval shape than undergoes the fourth moult and emerges as a full grown adult. After mating the female starts egg laying within the body and dies after sometimes with the result body is converted into a cyst. Only one generation is found in a year.

Control

1. Crop rotation is the most practical method ; cereals may be replaced by pulses.
2. Use of DD fumigant @ 225 lit or D B C P @ 20 lit/hec found very effective to control the nematode.
3. Application of aldicarb 10 gm @ 15 to 20 kg or carbofuran 3 gm @ 30 kg per hectare has also found effective.

Root-knot Nematode

This was the first plant parasitic nematode reported by Barber in 1901 from India. The root-knot nematode belonging to the genus *Meloidogyne* is most widely distributed and attacks almost all growing plants in every corner of the country. There are several species of this genus out of which following are more common:

Meloidogyne javanica, M. incognita, M. hapla, M.arenaria and *graminicola.*

Symptoms: The important symptoms are given below:

1. The infected plants are stunted and poor in growth.
2. The leaves exhibited discolouration and smalling in size.
3. The plants show reduced, unthrifty growth and a tendency to wilt during warm days.
4. When population is very high, the young seedlings may be killed.
5. The growth of roots is affected, and finally galls are produced. Sometimes, the galls also appear on the other parts of the plants.
6. The flowering and fruiting are delayed.

Life History

The gravid females of the nematode are found in the tissues of galls which die after laying the eggs in the clustures outside the body. The eggs are covered with a gelatinous membrane.

A cluster may contain 250-300 eggs which are oval in shape. The eggs hatch under favourable conditions and second stage larva is liberated free into soil. The larvae may enter from any part of the root preferably near the root tip. They are sedentary parasite and once they have become located within the tissues of the plant they do not again move to change the position. After living as a parasite for about three weeks, a male moults 3 times in rapid succession and undergoes a metamorphosis from which it emerges as a slender worm.

The female undergoes the same moults as the male and at about the same time. A female continues to grow, increasing in girth and to some extent in length until it becomes pear-shaped. If the plant is suitable host and the weather is warm, females begin to lay eggs from 20-30 days after penetrating the roots as a larvae.

Control

1. DD fumigant @ 200 lit or D B C P @ 25 lit per hectare found effective.
2. Aldicarb 10 gm @ 20 kg or carbofuran 3 gm @ 30 kg per hectare has been quite effective in reducing the nematode population.
3. Fallowing of field for some times, hot weather cultivation and cultural practices are also effective for reducing the population.

Some Nematicides

DD: It is mixture of 1, 3,-dichloropropene 1, 2 dichloropropane. It is used as a soil fumigant and is a colourless to brownish liquid. DD is slightly soluble in water but soluble in organic solvents. It may injure the germination, if the seeds are sown within 2 to 3 weeks after treatments. DD is generally applied to control the plant parasitic nematodes @ 200 to 400 lit/hec against various species of plant parasitic nematodes.

Vapam: Is also known as sodium methyl di-thiocarbamate. Vapam is soil fumigant with excellent nematicidal properties and is widely used on non-edible crops. It is white crystalline salt readily soluble in water. Vapam is formulated as 31% solution in water and used generally @ 30 lit./hec.

VC-13: It contains 80% of an organic phosphorus compound that is an effective nematicide. The active compound is 0, 0 diethyl, 0-2-4-dichlorophenyl phosphorothioate. It is less volatile and non phyto-toxic. VC-13 can be safely applied to living plants as drench @ 40-50 lits/hec.

The following systemic granular chemicals have also been found as effective nematicides and widely used for the control of nematodes.

(a) Aldicarb 10% gm. (b) Phorate 10% gm.

(c) Carbofuran 3% gm. (d) Disulfoton 5% gm.

(e) Fensulfothion 5% (f) Nemacur 10% gm.

Rat: Rats are gnawing mammals belonging to Order Rodentia. They are unwelcome associates of mankind from time immemorial. They cause a colossal loss all over the world. It is estimated that these animals destroy 30% of the total world's cereal production. In India, 10 million tonnes valued of rupees 100 crores are lost every year (Sood and Gurya 1976). On the basis of the estimates of losses inflicted by rats, it is often claimed that if rodent's ravages could be controlled, there would be no need of importing food grains from foreign countries. Besides, there has been enormous losses of human lives due to many diseases spread by these abnoxius animals. Rats and mice being prolific breeders, their population keeps on increasing very rapidly. According tc Devras the rat population in India is to be 2400 million, while Srivastava (1966) claimed it to be 48 millions.

Habit

Rats are very cunning and clever animal. They are very much afraid of new object which comes in their way. They are colour blind. They have well developed senses of smell and taste and possess good hearing capacity. They bear chisel like teeth. They generally move for the search of food in the night. They are good swimmer, they can climb pipes, walls and trees and can jump a horizontal distance of about 8 feet or can drop from a height up to 14 feet.

Habitat

Field rats dig their burrows inside the dense vigorous crops and house rats near sheltered places close to food storage. There are chambers inside the burrows which are used for the purpose of breeding and storage of the food materials.

Life Cycle

Rats breed through out the year and giving birth from 6-12 young ones in each year. Some times number of young ones may be 20-22. Four generations are found in a year. Females become capable of reproducing after three months of their birth. The offsprings open their eyes in 15 days and ears in 2 days. It is estimated that a pair of rats breed in year so rapidly that their number sometime becomes 1000 to 1500. The life span of rat generally varies from 1-3 year.

Some Common Species of Rats

On the basis of their shelter rats can broadly be divided in the two groups.

1. House rats
2. Field rats.

House Rats: Following species are main in this group: 1 *Rattus rattus* (Black-rat) 2. *Mus musculus* (House mouse)

Field Rats: This group comprises following species of rats:

1. *Bandicota bengalensis (lesser bandicoot)*
2. *Tatera indica (Indian gerbil)*
3. *Nesokia indica (Short tail mole rat)*
4. *Rattus meltada (Soft fur field rat)*
5. *Mus booduga (Field mouse)*

Common species of rats found in both field as well as houses.

1. *Rattus norvegicus (Norway rat)*
2. *Rattus rattus* rufescens
3. *Bandicota indica (Large bandicoots)*

House Rat **(Rattus Rattus)**

This species is known to cause great damage to food grains and domestic articles by dwelling in houses, warehouses, packing houses, seed stores, ships and sea ports. House rat consumes 10 gm food grains per day and exterminates much more. They are mostly found in human habitted places. They are black or grey above and pure white or pale lemon belly. Tail is often larger than body length. The body weight of adult rats ranges from 300-350 gm. It is essentially an arboreal or climbing animal and burrows.

House Mouse **(Mus Musculus)**

It is very active animal. It feeds 2-5 gm food grains per day. It is generally found in house but some times in gardens and fields near the villages and towns. The tail is larger than the body length. The colour of the body is dark to light brown above and pale from below.

Lesser Bandicoot **(Bandicota bengalensis)**

This rat is dark brown above and pale brownish below. The tail is short about three quarter of the head and body length. It is a robust field rat. They mostly' feed on crops vegetables, and fruits. Generally one rat is found in each burrow. They are good swimmers.

Indian Gerbil **(Tatera indica)**

The body colour is light brownish or sandy on the upper surface and darker with pale haired in the lower-side. The tail is longer than the head and body length. The eyes are very large. It mainly feeds on grains, succulant stems, leaves, roots, fruits, grasses and insects. More than one rat is usually found in a burrow. It is very poor in swimming. It makes its burrows in the dry land patches near the field, river bands, village site or under bushes etc. This rat does not hoard any food material in his burrow.

Rat Control

Rat control is a very complicated problem. It needs a thorough approach to minimise the rat menace. Some methods are given below to deal with the problem.

Rat Proofing: It checks the movements of house rats and mice into or outside the house, farm, storage and other building. For this purpose, all smallest and larger openings are completely, sealed and closed. The edges of all doors, windows and other inlets and outlets are covered with metal sheets. Concrete walls and shallow foundations should be constructed. Devices should be installed to keep the door closed. Food grains and other materials may be kept in metallic container or boxes.

Cleanliness: The heaps of trash, garbage and sweepings should not be allowed for a longer period near the houses and godowns. The food materials should not be kept in the open containers. It is very essential to launch clean up programme in the rat infested areas so that rats may not harbour themselves in the want of food and shelter. In this way rat population can be minimised.

Traps: The traps are employed in the places where rat infestation is small and chemical control like poison baiting. Fumigation is impracticable. There are many types of traps, *viz.*, cage traps, wooden traps, wire traps, snap traps and bamboo traps which are commonly used in the country. Wonder traps manufactured at Jalgaoin, Maharashtra by galvanised wires based on the designs recommended by Hoffkins Institute Bombay, *viz.*, bandicoot trap, double mouth sure success trap and union trap etc, are more efficient, for trapping of rats. The traps should be cleaned by water after each installation and fresh and attractive bait material like breads, fruits etc. should be used.

Poison Baiting and Fumigants: Poison baiting is generally used to kill the rats with various rodenticides. The rodenticides can broadly be divided into two groups.

(a) *Single dose Poison:* This group comprises acute poison which kill the rats by single exposure to poison bait. These are zinc phosphide, barium carbonate, noirbormide arseniozide, sodium fluoro-acetate etc. Out of these, zinc phosphide is commonly used.

(b) *Multiple dose Poison:* These, chemicals are effective only after the multiple feeding of poison baits by rats, because their cumulative effects leads the death of animals. This group includes warfarin, ratafin, fumarin, rodafrin, etc. Out of these ; the warfarin is commonly used. Being less poisonous, they are safer to mankind and domestic animals.

Some Newer Rodenticides

Vacor (RH-787), Silmurin and Bromadiolone are of very recent origin. They are found to be most effective against rats. Their single dose of poison baits kills the rats. No bait shyness reflex build up by the use of these poison. They are safer to human beings and domestic animals.

Preparation of Poison Bait with Zinc-phosphide

2% poison bait kills the rats. One part of zinc phosphide is mixed with 49 parts of grains. The edible oils like mustard oil etc. are added in small quantities to make a complete and attractive bait. 10-15 gm. poison bait is kept inside the burrows with the help of spoon etc, and is closed gently by putting grains and soil clods on the openings. Since zinc-phosphide is very poisonous hence, the bait should be prepared under technical care and be kept beyond the approach of children and domestic animals.

Preparation of Poison Bait with Warfarin

Following ingredients are used for the preparation of poison baits:

1. Warfarin-25 gm.
2. Grounded grain-450 gm.
3. Sugar-15 gm
4. Edible oil-10 gm.

These are mixed thoroughly and kept in the earthen vessels. These vessels are kept in different places in the houses, and godowns etc. After several days of feeding the rats die without postulating any sudden alarm of danger to their community.

The newer rodenticides like vacor (RH-787), Silmurin and Broma-dioloine are used as 1% and 0.05% poison bait.

Fumigants

Some chemicals produce gases which are poisonous and kill the rats. The commonly used chemicals are cynogas dust, hydrocyanic acid, celphos or phosphume tablets. 0.6 to 0.8 gm. celphos or phosphume tablets are used in each burrow. The rat burrows are made air tight for the application of fumigants. All the openings should very well be closed, after detecting them by passing air in the burrows through rat fumigating pumps.

Chemosterilants: The chemical which makes the rat sterile is called chemosterilant. Furadantine and colchicine are used as rat chemosterilants 0.02 gm. furadantine and 0.14 gm. colchicine, respectively make the male and female rat sterile. These are generally used as mixture of one tablet of furadantine and 1/2 tablet of colchicine in the wheat flour to make both male and female sterile.

Integrated Technique to Control Rats

An integrated technique has been evolved, developed and standardised in the Department of Entomology at C. S. Azad University of Agriculture and Technology, Kanpur which successfully exterminates the rats.

The plan of operation is given below:

(i) First day-survery and closing of rat burrows.

(ii) Second day-prebaiting in the opened burrows.

(iii) Third day-prebating continued.

(iv) Fourth day-2% poison baiting with zinc-phosphide.

(v) Fifth day-fumigation by celphos or phosphume tablet at the rate of 0.3 gm. per burrow. The prebaiting is practised to make the rats habitual of feeding baits provide them.

National Programme for Rodent Pest Management

A national programme for Rodent Pest Management has been drawn by the Indian Council of Agricultural Research. The plan has been drawn with a view to involve the entire community to undertake the campaign on nation wide bases. The programme has been divided into 3 phases that is: (i) Trainers training, (ii) Preparation of the community, and (iii) The rodent control strategy which involves the control operations at crop fields, threshing floors and residential houses followed by evaluation and sustenance strategy. Under this programme sofar training is being imparted at seven regional centres located at Jodhpur, Hyderabad, Parbhani, Cuttack, Shillong, Hapur and Mysore. In order to create awareness in the public and involve their active association in the rodent control campaign publicity material has been prepared and distributed.

The whole programme is depicted in the following flow chart:

Controlling the Rodents

Phase I. Trainer Training Student Squad Jan-Feb. Village Volunteers and Youth Club

Phase II. Preparation of the community. All India Radio T.V. Site Programme, Field Publicity Units, Directorate of Extension and Universities. March-June

Phase III. Rodent control strategy control operation

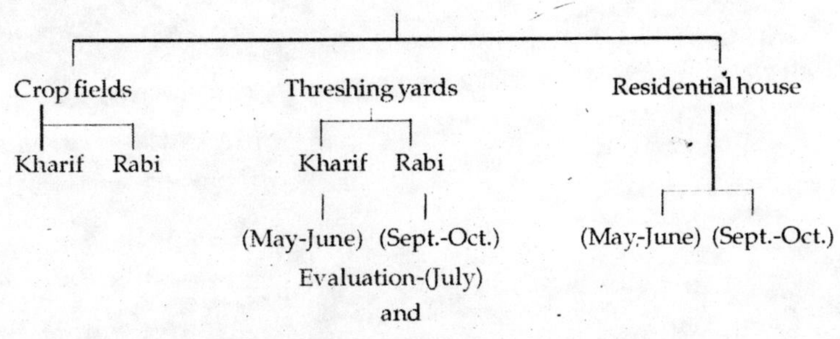
Crop fields
Threshing yards
Residential house
Kharif
Rabi
Kharif
Rabi
(May-June)
(Sept.-Oct.)
(May-June)
(Sept.-Oct.)
Evaluation-(July)
and
Sustence strategy (As and when required)

2

The Fundamentals

The term entomology has been derived from the Greek words, entomon = insect and logos = knowledge, thus in broad sense it may be defined as a branch of Zoology which deals with the insects. Amongst 20 phylums of animal kingdom, Arthropoda is the biggest one and comprises the class Insecta, which is the largest and complicated group of insects.

It is estimated that there are more than 13, lakhs animal species exist in the world and out of these species nearly 9 lakhs fall under the class Insecta. Therefore, to make the study of insects more systematic and simpler, the subject has been divided into the following important branches.

Morphology

It deals with the study of structure of insects and has the following sub-branches:

(a) *Anutomy:* The internal organs are studied under this head.

(b) *Histology:* It includes the study of tissues under the microscope.

(c) *Cytology:* This is the science of cells and their constituents.

Taxonomy

This branch includes the classification of insects and rules of their nomenclature.

Embryology

It deals with the growth and development of embryo.

Physiology

This branch principally deals with the functions of organs, cells and the tissues.

Ecology

It includes the study of environment in relation to insects.

From the economic point of view, the following are the important branches of Entomology:

Agricultural Entomology: This mainly concerns with the study of insects which are directly related with the crops and the stored commodities.

Medical Entomology: This branch deals with the insects, which are responsible for transmitting human diseases.

Veterinary Entomology: It includes the study of those insects, which may be beneficial or harmful to the animals.

Industrial Entomology: This branch is concerned with the insects, which are directly or indirectly related with wood, wool, cloth and the other industrial materials. It also includes the study of beneficial insects like honey bee, lac insect and silkworm.

Forest Entomology: It includes the study of insects concerning with forest plantation.

Paleo Entomology: Under this branch the studies are made on fossils of insects.

Phylum Arthropoda and its Classes: Arthropoda (Gr. arthros = jointed, podo = limbs)

This is the largest phylum (Snodgrass, 1952) of the animal kingdom. The main characters of this phylum are given below:

1. Bilaterally symmetrical animals.
2. They are triploblastic (with 3 germ layers), *i.e.*, ectoderm, mesoderm and endoderm.
3. Metameric segmentation, *i.e.*, body of the animal is divided into a succession of homologous parts of metameres.
4. Segments fused to form head, thorax and abdomen.
5. Majority of the segments possess pair of jointed appendages.
6. Exoskeleton contains chitin secreted by epidermis.
7. Gills and muscles are emphasized, therefore, rapid movements are possible.
8. Striated muscles are emphasized, therefore, rapid movements are possible.
9. Specialized alimentary canal with chitinous teeth, chambers and gastric ossicles present.
10. Metamorphosis takes place in the development.

This phylum is divided into seven classes:

1. Onychophora 2. Diplopoda 3. Chilopoda 4. Symphyla 5. Crustacea 6. Arachnida 7. Insecta

ONYCHOPHORA

It is a group of primitive arthropods represented by a single genus peripatus. The important characters are as follows:

1. The body is soft, worm like and not divided distinctly into regions.
2. They have short, stumpy unjointed appendages.
3. The excretory organs take the form of metamerically repeated coelomoducts.
4. They possess tracheae as respiratory organs.

Example: Peripatus.

DIPLOPODA

The important characters are as follows:

1. They have the greater number of body segments and body is cylindrical.
2. They live in humid dark places under stones or beneath or within rotting logs and avoid light.
3. They have a pair of stink glands in most of their body segments except some anterior and some posterior segments.
4. Each segment carries two pairs of legs and two pairs of spiracles.
5. The gonads open behind the 2nd pair of legs.

Example: Millipedes.

CHILOPODA

It possesses following characters:

1. These are slender elongate, segmented and flattened dorsoventrally.
2. They are usually provided with a single pair of appendages and a pair of spiracles to each of the post cephalic segments.
3. The first pair of legs is modified to form poison claws.
4. The gonads open on the penultimate segment of the abdomen.
5. Head possesses one pair of antennae, one pair of jaws and two pairs of maxillae. *Example:* Centipedes.

Symphyla

The major characters are as under:

1. They have long antennae and most of the body segments possess a single pair of legs.
2. White eyeless animals, less than 10 mm long.
3. There is a single pair of spiracles which is situated on the head.
4. The gonads open on the 4th post cephalic segment.

5. They have mimcrous legs and live in soil and avoid light.

Example: Garden centipede.

Crustacea

The important characters are described below:

1. They are characterised by the presence of two pairs of antennae and at least 5 pairs of legs.
2. Their body is covered externally with chitinous cuticle and the exoskeleton, which consists of a series of hard sclerites, is separated by soft flexible chitin.
3. Six anterior segments are fused with the pre-segmented region to form the head, the rest are thorax and abdomen. Sometimes thoracic segments are fused with head to form cephalothorax.
4. Respiration takes place by means of gills.
5. The excretory organs are generally modified coelomoducts, which are represented by green glands or shell glands.
6. The genital apertures are situated anteriorly.
7. Many crustacians possess protective and/or aggressive characters.

Examples: Crabs, lobsters, shrimps.

Arachnida

They are air breathing arthropods and possess following important characters:

1. The body is usually divided into cephalothorax and abdomen.
2. There are four pairs of legs and antennae are absent.
3. The excretory organs are generally malpighian tubules.
4. The gonads open near the base of abdomen.
5. In place of antennae chelicerae are found.

6. Respiration takes place by means of bronchiae which form lung books, or by tracheae or spiracles.

Examples: Scorpions, king crabs spiders, mites and ticks.

Insecta or Hexapoda

It is the largest class of the phylum and the important characters are as follows:

1. They are tracheate arthropods.
2. Body is divided into head, thorax and abdomen.
3. Breathing takes place through direct tracheae and tracheoles.
4. Head bears a pair of antennae, thorax bears 3 pairs of legs, usually one or two pairs of wings in the adult stage.
5. The development is rarely direct and generally it is indirect (metamorphosis).

Example: Insects. (Housefly, grasshoppers)

Insects and their Characteristics: Insects are tracheate arthropods (Greek Arthros = segmented, podos = leg) in which the body is segmented with a relatively tough integument and jointed limbs.

The origin of arthropods is from annelid-like ancestor with the body bearing a series of ring like segments; a simple head and ventrally placed mouth. In the animal kingdom, phylum Arthropoda includes five classes of animals *viz.; Crustacea* (crabs, prawns and lobsters), *Onychophora* (peripatus), *Myriapoda* (mellipedes, centipedes,. *Insecta* or *Hexapoda* (insects) and *Arachnida* (spider, scorpion, ticks and mites).

The main characteristics of insects (*Latin insectum = having cut into)* are summarised below:

1. The body is bilaterally symmetrical *(i.e.,* the longitudinal half is the mirror image of another half) and triploblastic (presence of three germ bands during embryonic development *viz.,* ectoderm, mesoderm and endoderm).

2. The body is divided into three distinct regions the head, thorax and abdomen.
3. The exoskeleton or the integument is hard and the flexibility is because of the chitin present, thus the integument combines strength and rigidity with flexibility.
4. The head is formed by the fusion of six embryonic segments.
5. The anterior region of head is provided with the mouth parts adapted for biting and chewing, piercing and sucking, siphoning, sponging and chewing and lapping etc.
6. There is a pair of segmented antennae situated near the eyes except the order Protura in which they are absent.
7. The thorax consists of 3 segments *viz.*, pro-, meso- and meta- thorax having one pair of legs in each segment. The second and third thoracic segments carry a pair of wings. Usually there are two pairs of wings, but some times only one pair is functioning and the other is reduced, vestigial or absent.
8. The abdomen is comprised of 7 to 11 segments.
9. The alimentary canal is tubular in structure which opens anteriorly into the mouth and ends into the anus. In place of liver, there is a pair of salivary glands extending along the sides of alimentary tract and opens into the mouth.
10. Circulatory system is open type *i.e.*, the blood is circulated in the body through the lateral openings of the heart known as ostia.
11. The blood is devoid of red blood corpuscles and yellowish green in colour.
12. The respiratory system consists of tracheae which open exteriorly by means of paired spiracles.
13. The nervous system is composed of the ganglionic masses which are interconnected with nerve cord.

14. Excretion takes place by means of malpighian tubules, nephrocytes, oenocytes and the integument etc.
15. Reproduction is usually sexual, sometimes viviparous and parthenogenetic.
16. Sexes are separate, but in some insects the male and female reproductive organs are present in the same individual like cottony cushion scale *(Icerya purchasi).*
17. Metamorphosis is a common phenomenon in insects.
18. Sense organs, which are usually present in insects include compound eyes, auditory (organs of hearing), olfactory (organs of smell), tactile (organs for sense of touch), gustatory (taste organs), chordotonal (sound producing organs) and the light producing organs.

Relationships of Insects: The relationships of the insects with the other arthropod groups have been the subject of much discussion and widely differing views have been proposed. The insects come to very near with the Crustacea and myriapoda by having mandibles and antennae and agree with the myriapods in having only a single pair of antennae, corresponding to the antennules of the Crustacea. In the absence of any direct evidence from palaeontology, our conception of the ancestry of insects has to based upon whatever data can be seen from embryology and comparative morphology.

It is generally conceded that the Diplura and Thysanura are the most primitive insects. The Diplura are nearest related, through the symphyla, with the ancestral stock from which the insects were presumably derived.

The Thyranura, on the other hand, combine Dipluran features with those of the lower pterygota. Their ectognathous mouth parts, annulated antennae, segmented torsi, together with the possession of compound eyes and ocelli, are characters which are shared with all of the more generalised winged orders. Furthermore, their general body-form prefigures that of the nymphs of these same orders. Reviewing the Arthropoda as a whole, it will be noted that:

1. The Onychophora are far removed from other Arthropoda since they show lower degree of development of these special feature that characterise the phylum as a whole. They rank more or less as a halfway class between the Arthropoda and annelidan ancestors.
2. The main evolutionary stem of the arthropods is represented by the Crustacea-Insecta-Myriapoda series which is characterised by a 6 segmented head (except some Myriapoda) bearing antennae, mandibles and maxillae. The Trilobita also have been placed with them.
3. The Arachnida have no differentiated head, the antennae are replaced by chelicerae and neither mandibles nor maxillae are present. It will be evident, therefore, that they represent a separate line of arthropod evolution and are not closely related to any of the other class.

3

Protection from Insects

The term plant protection refers to the measures which are used to protect the crop from pests and diseases. Plant protection like any other scientific discipline embarrasses in its fold the body of knowledge, collected through assiduous and careful observations over the years for raising a pest and disease free crop and protecting the produce during transport and subsequent processing.

The term pest broadly includes all living organisms enemical to the interests of man such as plant pathogens, weeds, harmful insects, mites, nematodes and mammals like rodents, bats, jackals, monkeys, elephants etc. The pest problem perhaps arose out of the necessity of man of olden days for growing plants of his choice over large areas and starting agriculture. Such practices upset the biological equilibrium though it helped to build up modern civilization. Such changes in the eco-system provided insect pests and plant pathogens with ready and available food supply. This battle between the man and foes of his plants has been going on for centuries now. The ingenuity of man's nature has developed a plethora of methodology and weaponry to fight these battles. The power of adaptability of his force is, however, boundless. It is possible to annihilate an insect pest or disease completely. Efforts should, therefore, be directed towards suppressing the pests and diseases and keeping the loss within economically safe limits.

It has been conservatively estimated that the losses inflected by pests to the food grains are the tune of 30% at the global level. In India about 20% of the crop yield is lost due to the ravages of insects-pests and diseases which approximately amount to rupees 1500 crores. Some times the epidemics of pests and diseases result in the complete failure of the crop and the striking example is of a famine due to the epidemic of a fungal disease caused by *Helminthosporium* on paddy crop in West Bengal during 1942-43. Bindra 1975 reported that the losses caused by insects, rodents, microbes etc. during storage, are as high as 25 to 30 per cent. Keeping in view the magnitude of losses caused by the pests and diseases, it is imperative to work out their effective control measures and to popularise their technical know how at the grass root level.

Importance and History of Plan Protection in India: The importance of plant protection in agriculture based country like India needs no emphasis, firstly, because of the fact that the conditions for the multiplication of pests and diseases are very conducive, particularly, during rainy season and secondly due to the ignorance of modern approaches of insect control. International food and agriculture organisation has stated that the destruction of crop pests is a subject of prime importance in Agriculture production. Not only this, in 1945. Wood head famine enquiry commission has suggested in the report that to take the maximum advantage of irrigation, manures and fertilizers and improved seeds to boost the agricultural production, we shall have to combat crop enemies with technical skill by using effective methods of control. The Eminient Entomologist late Dr. Pradhan has also emphasised that Protection Technology is more important than Production Technology.

Inspite of the above facts unfortunately we do not have the accurate and precise information with regard to the losses caused by pests to different crops. However, George Oridish (1961) estimated the losses caused by insect-pests to different crops amounting Rs. 485 crores annually and the details are as follows:

S.N.	*Crop*	*Losses in rupees crores*	*Number of insects involved*
1.	Paddy	1.37	6
2.	Other cereals	112.47	6
3.	Pulses	44.94	3
4.	Potato	2.90	4
5.	Sugarcane	52.98	3
6.	Cotton	29.17	6
7.	Fruits and vegetables	61.20	a number of insects
8.	Other crops	44.47	— do —

Since a fore said data are about 40 years old and now the ravages of pests and diseases increased enormously due to the modern agronomical practices and high yielding varieties, it is therefore, necessary to work out the losses caused by them like other developed countries. This information will help us to work out the effective plant protection measures for particular crop.

Brief History

It is needless to say that the heavy magnitude of losses caused by insect-pests have attracted the attention of administrator to initiate the work on plant protection. No doubt we get the mention of lac insect, mites, bugs moths in our ancient literature but their relation can not be linked with the scientific study. This was Koening who initiated the studies on insect pest as early as 1799 in South India and he discovered about 100 insect species from that region.

Thereafter important contributions during this period were made on lac insect in 1782, 1790 and in 1800. With the growing realization of the importance of insect pests an Asiatic Society was established in Bengal during 1785 and this society encouraged the collection, exhibition and its publications contained valuable contributions on Indian insects. This society generated the awareness so much so that in 1875 an Indian Museum was founded at Calcutta and collections of insect species made by the society were transferred to this museum.

Later in 1883 Bombay Natural History Society was established and a first research journal was published by this society in 1888. The perusal of literature reveals that the practical work on Economic Entomology got the momentum by the end of eighteenth century and as per advice of Sir Advert (revenue secretary), five volumes on Economic Entomology were published during 1889 to 1903 from Indian Museum.

The first Entomologist, Lionel Niceville, who came to the Government of India was appointed in 1901 and posted at Indian Museum Calcutta. But after his sudden death, Maxwell Lefroy was appointed on this post in 1903. These individual appointments were soon followed by the establishment of Imperial Agricultural Research Institute in 1905 which is now known as Indian Agricultural Research Institute and Lefroy was appointed its Director. He wrote two books *viz.*, Indian Insect Pests and Indian Insect Life in 1906 which are monumental in the field of Entomoloty.

In the same year Forest Research Institute was started at Dehradun which has a separate well developed section of Entomology. The year 1916 saw the birth of yet another important organization the Zoological Survey of India at Calcutta. In the beginning of 19th century the entomological work of agricultural importance was initiated as early as 1902-03 in Madras, 1905 in Punjab and 1906 in U.P. To undertake and coordinate the research work in these states separate Entomlogists namely Coleman at Mysore in 1906, Fletcher at Madras in 1912, Afzal Husain in Punjab in 1913 and Richards in U.P. in 1922 were appointed. In 1921, Central Cotton Committee was constituted to finance various entomological schemes. On the recommendation of Lac commission (1921). The Indian Lac Research Institute was established at Ranchi (Nankum) in 1925.

In order to coordinate the various activities at the different agricultural centres established in different parts of India, Indian, Council of Agriculture Research was founded at New Delhi in 1929. It also provides the technical and financial aid to these centres.

The real history of Plant Protection begins with the enforcement of Destructive Insects and Pests act of 1914 which also includes the Plant Quarantine Act. Looking to the importance of Plant Protection many State Governments provided facilities and staff for this important work and fully established and organised Plant Protection Centres as integral parts of the departments of agriculture. To have the over all supervision in the field of Plant Protection in the Central Directorate of Plant Protection, Quarantine and Storage was established in 1946 which is now located at Faridabad.

Legal Protection

In order to check the entry of insect pests and diseases from one country to another, it was felt necessary to enforce the plant quarantine act in 1912 so as to prohibit the importation or shipment, interstate of any class of plants or plant products from any foreign country or locality and from any stage or portion of state or territory in the country. Such specific prohibitions are called quarantines. At the international forum the plant quarantine acts namely foreign quarantine and domestic quarantine acts were enforced. At present foreign plant quarantines include the restriction of importation of certain fruits and vegetables; cotton lint and cotton seed products, certain cereals, plants and plant products for propagation or as packing materials. Under domestic quarantine act the State Governments can send food grains and vegetables to other state only after obtaining phytosanitary certificate.

Such provisions have also been made for the import or export of these items. Realising its importance, Indian government has established a centre at Bombay sea-port in 1949 and at Madras 1950. There after in 1955, 56 and 57 such stations were also started at Cochin, Calcutta and Vishakhapatnam, respectively. At all these centres plant quarantine laboratories were developed which verify the certificate issued by government whether the food grains or the other materials is free from the pests and diseases or not. In case the material is found infested, it is fumigated at that very station. The cotton which is to be imported is invariably fumigated at these stations, so that the cotton boll weevil may not get the entry in India.

Like sea-ports, such stations have also been established at the air ports for Amritsar, Delhi, Madras, Bombay and Calcutta. In 1951 India has also become a member of International Plant Protection Organisation. This organisation meets from time to time to discuss the various effective methods of plant protection. After such meetings an International Managing Committee was constituted which organised the Ist conference at Bangkok in 1956 and IInd, IIIrd at Ceylone and Delhi in 1957 under the Plant Quarantine Act. Indian Council of Agriculture Research sanctioned a scheme on the survey of the breeding places and life cycle of desert locust in 1930.

Locust Warning and Central Organisation: After the completion of work on locust scheme sanctioned in 1930, Indian government established a permanent locust warning and control organisation under ICAR in 1939. Dr. Hem Singh Pruthi was the incharge of this organisation. Its head quarter was at Karanchi and six specialists were working in this organisation. In 1940 a locust swarm entered in the country, though organisation made its effort but could not get much success. Therefore, in 1942 a coordinated anti-locust scheme was started resulting into the formation of regional locust organisation in different states and the locust control was started with the help of regional officers.

At present Government of India is responsible for the desert locust control over an area of about 82,000 square miles in the desert area of four states and maintains a permanent Locust Warning Organisation This organisation's functions are to constantly survey desert area, to note and record incidence and fluctuations in locust populations, to observe their behaviour and habits in relation to conditions of climate, weather, vegetation and soil and to adopt control measures against them.

Organised Protection

It the centre, Plant protection Quarantine and Storage was established in 1946 at New Delhi, which is headed by the Plant Protection Advisor to the Government of India who is also the director of locust control. The present Plant Protection Advisor is Dr. V. Raghunathan. The important functions of this directorate are given as below:

1. The distribution of information within the country regarding the pests and diseases of plants and plant products and the means of their prevention and control.
2. To ensure the availability of pesticides arid the appliances to be used for this purpose.
3. To transfer the technical know how with regard to the control of insects, diseases and rodents infecting stored commodities.
4. Forecasting and warning of the out break of desert locust and to provide help for its control.
5. To provide technical and financial help for the outbreaks and epidermics of pests and diseases in states and union territories.
6. Implementation and the operation of the Central and International Plant Protection and Quarantine Act.
7. Survey for the national pests and disease campaigns for their control to organise the training programme for the officers of different states.

8. To depute government officers in foreign countries for the latest knowledge in the field of plant protection.
9. To provide the help to the central government with regard to follow the rules and regulations related with plant protection.
10. To launch the campaigns for the control of pests and disease under Grow More Food Campaign.
11. To provide help in procuring the technical material of pesticides from the foreign countries and the help in establishing the firms for manufacturing and formulating the different pesticides.

Organizational Setup of Plant Protection

The Directorate of Plant protection, Quarantine and Storage Works under the Ministry of Food and Agriculture, Government of India. The Plant Protection Advisor and Director Locust Control is the head of this organization.

The growth and development of plant protection organisation has been depicted in the following table:

Period	*Man Officer*	*Power Staff*	*Expenditure (Rs. in Lakhs)*
1946	12	117	7.05
1950-51	18	449	14.08
1955-56	26	383	31.84
1960-61	53	656	63.83
1965-66	67	734	53.83
1970-71	59	701	61.22
1975-76	112	933	125.38
1980-81	179	1223	321.05
1985-86	198	1554	717.64
1988-89	217	1617	773.61

Different Divisions/Units of Plant Protection

1. Integrated Pest Management
   1.1 Pest and diseases surveillance
   1.2 Biological control
2. Locust warning and control
3. Plant Quarantine
4. Pesticides regulatory measures
   4.1 Secretariat of Registration Committee/Central Insecticides Board.
   4.2 Central Insecticides laboratory.
   4.3 Regional Pesticide Testing laboratories.
   4.4 Pesticides Monitoring Unit.
5. Plant Protection Training.
6. Documentation.
7. Centrally sponsored scheme for control of pests and diseases.

Systematic Control

Pest and Disease Surveillance

The following are the main objectives:

(a) to monitor and forewarn the likely pest and disease build up on crops.

(b) to popularise Integrated Pest Management (IPM) approach among extension functionaries and farmers.

(c) to minimise environmental pollution due to pesticides.

(d) to maintain ecological balance of natural fauna through judicious and safe use of pesticides. The major functions of this unit are as follows:

  (i) to assist State Government in pest survellance on major crops.

  (ii) to supplement state efforts in pest control operations.

(iii) to monitor and forewarn pest and disease situation in the country.

(iv) to disseminate IPM concept.

(v) to advise appropriate and safe use of pesticides.

Activities and Functions

1. Conduct sample Rapid Roving Pest Surveys on crops and generate field data on pest/disease incidence through fixed pest observations and light trap studies.
2. Issue pest situation and forewarning reports to state Departments of Agriculture.
3. Layout demonstrations on IPM technology in farmers' fields.
4. Organise training programmes on latest plant protection techniques/methods (IPM) for farmers and state extension agencies.
5. Assist state authorities in organising pest control programmes/campaign.

Set up and Infrastructure

All units work under the control of plant Protection Adviser to the Government of India. Besides, following staff are in position under I.P.M.

Director (IPM)

Joint Director (PP)

Deputy Director (PP)	Deputy Director (PP)
C.S. Stations (19)	C.P.P. Stations (13)
(Central surveillance)	(Central Plant Protection)
PPO (1)	PPO (1)
APPO (1)	TOT (2)
T.O. III (6)	
8	3

Biological Control Unit

Objectives:

(i) Control of insect pests and weeds by using natural enemies (parasites, predators and pathogens)

(ii) promotion of biological control as major component of IPM.

Functions:

(i) Conservation of natural enemies.

(ii) Augmentation of natural enemies

(a) Introduction and colonisation of exotic and indigenous natural enemies.

(b) Mass-multiplication of potential natural enemies in the laboratory for releases.

Activities:

(i) Survey of crops to determine the pest status.

(ii) Identification of potential natural enemies.

(iii) mass multiplication of potential natural enemies.

(iv) Conservation of natural enemies by assessing their control potential against pests to delay/avoid the use of pesticides.

(v) Educate the farmers regarding the identification of natural enemies and their role in pest control.

(vi) Supply of natural enemies to farmers and other agencies for pest control.

(vii) Advice States/ UTS on biological control of crop pests and Weeds and IPM programmes.

Setup and Infrastructure

11 Central biological control stations and a parasite multiplication unit is located in 11 states. Central Parasite breading laboratories at Gorakhpur and Banglore and Satellite Bio-control field Laboratories at Kullar, Thanadher and Vyara are proposed. Each station has normally 14-17 staff members and PMU Banglore has 9 staff members which is as follow:

PPA to the Govt. of India Joint Director (Entomology)

Faridabad headquarters

TO-III (1)

P.M.U. Bangalore (Parasite Multiplication unit)		C.B.C.S. (11 numbers)		
Deputy Director (Ento.)	-1	Assist. Director	1	
PPO	-1	APPO	1	
T.O. III	-2	T.O. III	1	each station
T.O. II	-2	T.O. II	5	
Contd...				
Lab. attendants	-1	Lab. attendant	1	
Field attendants	-1	Field attendant	1	

11 Centres of Central Biological Control Stations are located at Srinagar, Solan, Sri Ganganagar, Gorakhpur, Hardwar, Bhubaneshwar, Raipur, Surat, Hyderabad, Banglore and Faridabad.

Locust Warning and Control: Locust warning organisation (LWO) was set up in 1939. It was merged with Central Directorate of Plant Protection in 1946. Field stations for investigations on locust (FSIL) was established in 1957 at Bikaner.

The main objective of this unit is to protect crops from locusts.

Functions:

(i) Keep watch on locusts in their natural desert habit

(ii) Advise State Governments on locust situation.

(iii) Maintain laws with locust prone countries and F.A.O.

(iv) Take control measures in scheduled desert areas.

Activities:

(i) Conduct survey around strategic locations in scheduled desert areas.

(ii) Timely warning on locust situation to the States through fortnightly bulletins and special communications.

(iii) Training on locust to the State Officials.

(iv) Participation in international meetings on desert locusts.

(v) Organise locust control in scheduled desert area.

(vi) Conduct field investigations on locusts.

Setup and Organisation: PPA to the Govt. of India is a Director (Locust control)

Joint Director
Deputy Director

Transport Engineer Jodhpur	***Assistant Director Jodhpur***		***Deputy Director FSIL, Bikaner***
Jodhpur circle out posts	Bikaner circle out posts	Barmer circle out posts	Palampur circle out posts
1. Jodhpur	1. Bikaner	1. Barmer	1. Palampur
2. Shergarh	2. Nokha	2. Chohtan	2. Tharao
3. Phalodi	3. Kolayatji	3. Balotra	3. Bhuj
4. Nokh	4. Pungal	4. Gadra Road	4. Jalore
5. Rokharan	5. Lunkeransar	5. Jaisalmer	5. Sanchore
6. Nagpur	6. Ganganagar	6. Ramgarh	
7. Didwana	7. Suratgarh	7. Mohangarh	
	8. Anupgarh	8. Sam	
	9. Nohar		
	10. Churu		
	11. Rajgarh		
	12. Sardar Sahar		
	13. Sikar		
	14. Narnaul		

Zonal offices are located.

	Total staff
1. Technical	145
2. Mechanical	53
3. Ministerial	39
4. Group 'D'	70
	307

Plant Quarantine

Objectives:

(i) To prevent introduction of exotic pests inimical to Indian fauna and flora.

(ii) To assist the foreign governments to do an international obligation.

Functions:

(i) Enforcement of Plant Quarantine regulations, Destructive Insect Pest Act 1914 and the plants, fruits and seeds (Regulation of import into India) order 1984 (PFs order).

(ii) Regulation of imports of plants, seeds/fruits, soil, living insects, fungi etc.

(iii) Quarantine inspection/treatment and release found free from pests, or otherwise deport/destroy.

(iv) Issue of phytosanitary certificate for exported agricultural commodities under the International plant Protection Convention 1951, of the F.A.I.

Activities:

(i) Inspection and treatment of plants/plant materials imported for consumption/propagation to avoid introduction of exotic pests.

(ii) Inspection/treatment and issue of phytosanitary certificates for plants/plant materials meant for export.

(iii) Mandatory foundation of build quantity of imported fresh/dry fruits, cotton balls and other plants/plant materials meant for consumption.

(iv) Collection, compilation and publication of information on pests/diseases of plant quarantine significance and their management by field survey of crops, godowns, storehouses, etc., and literature survey and review.

Setup—26 plant quarantiene and fumigation stations are located at various international airports/seaports/lead frontiers and total sanctioned strength of the staff is 354 as given below:

Group	*Non-plan*	*Plan*	*Total*
A′	—	11	11
′B′ (Gazetted)	24	6	30
′B′ (Non-Gazetted)	2	8	10
′C′	130	101	231
′D′	61	11	72
Total	217	137	354

Pesticides Regulatory Measures

Secretariat of Registration Committees/Central Insecticide Board:

Objectives:

(i) Registration of insecticides for the purpose of indigenous manufacture and import.

(ii) Advice to the Central and the State Governments on technical matters.

Functions:

Assist Central Insecticides Board (CIB) and Registration Committee (RC) and the various sub-committees of Central Insecticides Board.

Activities:

(i) process applications for registration of insecticides.

(ii) Issue certificates of registration for insecticides approved by the Registration Committee.

(iii) Act as secretaries of Central Insecticides Board and Registration Commodity

(iv) Work out the approved uses of different groups of pesticides and their formulation.

(v) Scrutinise shelf-life claims in respect of pesticides.

(vi) Examine and recommend pesticides suitable for aerial application.

(vii) Deal with cases relating the inclusion of new pesticides in the schedule to the Insecticides Act, 1968.

(viii) Deal with court cases and other legal matters pertaining to Insecticides Act, 1968.

Setup and Organisation:

Secretary, CIB and RC—1— It has following units:

1. *Bio—efficacy Unit:* Principal Scientific officer-1

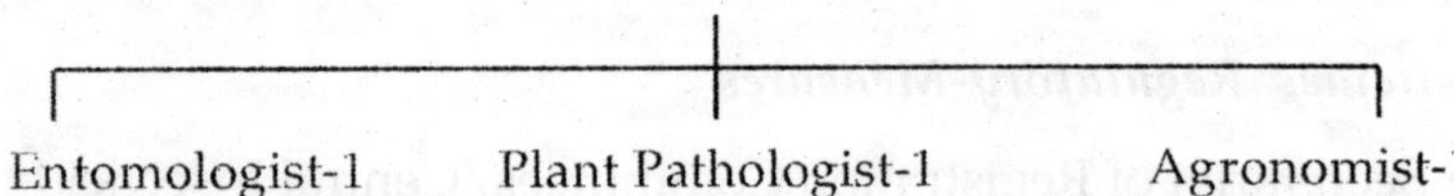

2. *Chemistry Unit:* Principal Scientific officer-1
   Senior Chemist-1
3. *Medical Toxicology Unit:* Principal Scientific officer-1
   Medical Toxicologist-1
4. *Packaging and Labelling Unit:* Principal Scientific officer-1
   packaging Engineer—1
5. *Legal Unit:* Law officer-1
6. *Administration:* Administrative officer-1

Central Insecticide Laboratory

Objectives:

Pre and post registration verification of the properties, performance and hazards of the product and the proposed use claimed by the manufacturer.

Functions:

(i) Verification of the quality of samples of insecticides, submitted by the officer or authority authorised by the Central or State Government.

(ii) Analysis of materials for insecticide residues.

(iii) Investigate insecticides with a view to verification of conditions of registration.

(iv) Determine the efficacy and toxicity of insecticides.

(v) Any other function as may be entrusted by the Central Government or by a State Government with the permission of the Central Government and after consultation with the board.

Organisational Setup

Multi disciplinary organisation with 4 defined areas, viz. Chemistry, bio-assay, packaging and medical toxicology.

Chemistry Division

Deputy Director-1

Engineer—1	Senior Scientific officer — 1	Junior Chemist—5
(Instrumentation)	Senior technical Asst. — 1	
Instrument	Technical Asst. — 1	
Technician-1	Lab. Technician—3	
	Lab. Assistant —10	
	Lab. Attendant-9	

Bio-assay Division

Deputy Director-1

SSO—1	SSO—1	SSO—2	SSO—1	SSO—1	JSO—1
Entomology	Nematology	Agronomy	Plant	Electron	
Radiotracer					
		Herbicide	Pathology	Microscopy	
JSO(Ent)—2	Tech.	JSO(Agr.	JSO (P1.	Tech.	
Tech.Asstt—1	•				
	Asst.—1	Herb.)—1	Path)—1	Asst.—1	
STA—1	L Tech.—2	Tech.	STA—1		
		Asstt.—2			
Tech.	L.Atten.—1	L.Tech.—l	Tech.		
Asst.—7			Asst.—2		
Lab. Asst.—2		Tractor	L. Atten.—l		
		Driver—1			
		Pump			
		Driver-1			
		Beldar-4			

Packaging Division

Deputy Director	—	1
Packaging Engineer	—	1
Junior Packaging Engineer	—	1
Technical Assistant	—	2
Lab. Assistant	—	3
Laboratory attendant	—	3
Mechanic	—	1
Helper	—	2

Medical Toxicology Division

Deputy Director-1

Medical Toxicology	Medical Pathology	Medical Biochemistry	Animal House
SSO—1	SSO—1	SSO—1	Veterinary—1 Incharge
STA—2	JSO—2	JSO—2 STA—2	Tech. Asst.—2 Animal house-2
Tech.Asstt.—2	Tech. Asst..—2	Tech. Asst..—2	Animal house Technician—6
Lab. Tech.—2 Laboratory	Lab.Tech.—2	Lab. Tech.—2	Animal house Attendant—7 attendant—5

Regional Pesticide Testing Laboratories

Objectives:

To supplement the resources of the State Governments and Union Territories Administrations in analysing the samples for monitoring quality.

Activities:

(i) Physico-chemical analysis of pesticides.

(ii) Providing co-ordination and guidance to States pesticides testing laboratories.

(iii) To assist in the formulation of Indian Standards for pesticides and their formulations.

Setup and Organisation

	Director (RL)
STA-2	Administrative officer—1
T. Assistant—6	Stenographer—1
Lab. Asst..—2	Store Keeper—1
Lab. Atten.—2	Peon / watch man—1

Pesticide Monitoring

Objectives:

Ensure pesticides availability for crop protection.

Functions:

(i) Assess States pesticides requirements for agricultural uses.

(ii) Monitor availability of pesticides.

(iii) Monitor implementation of 50% pesticides distribution scheme.

(iv) Accord permission to import samples of pesticides not included in schedule to the Insecticide Act, 1968, for evaluation.

Activities:

(i) Discuss demand with States and Industry.

(a) Co-ordinate with the Ministry of Industry and pesticide Industry for ensuring their timely availability.

(b) Recommend import as and when required.

(c) Formulate views on import policy.

(d) Advise States on the distribution.

(ii) Allocate insecticides on quarterly basis under the 50% distribution scheme to states for getting formulation through non-associated formulations.

(iii) Collect/compile data on pesticides production, import and distribution.

Organisational Setup

Central Directorate of plant protection is assisting union Department of Agriculture in ensuring availability of pesticides for crop protection.

Joint Director (E - II)

Plant Protection Officer

Plant Protection Training (Established in September 1966)

Objectives:

Human resource development in Plant Protection, in pesticide formulation analysis and pesticide residue analysis.

Functions:

(i) Training officials in Plant Protection.

(ii) Organise form for exchange of information.

Activities:

(I) Campus Training:

(a) Long term training

- (i) Post Graduate Diploma (10 months)
- (ii) Advance Plant Protection (3 months)
- (iii) Pesticide Formulation Analysis (3 months)
- (iv) Pesticide Residue Analysis (3 months)

(b) Short term training

- (i) IPM in rice, sugarcane and cotton (1 month)
- (ii) Plant Protection in Horticulture (1 month)
- (iii) Work shop-cum seminar for SDIs (3 weeks)
- (iv) Orientation for graduates of agricultural universities (5 days)

(II) In Situ Training in States:

- (i) Instrumental analysis of pesticides (10 days)

(ii) Audio visual aids (4 days)

(iii) Weed management (7 days)

(iv) Rodent control (7 days)

(v) Application techniques (4 days)

Documentation

Objectives:

To create awareness on Plant Protection technology.

Functions:

(i) Documentation of technical information on plant protection.

(ii) Publication of journal "Plant Protection Bulletin".

Activities:

(i) Collection, collation and dissemination of crop pest situation.

(ii) Preparation of periodical reports on various aspects of plant protection for submission to Department of Agriculture and cooperation and FAO.

(iii) Mass education on key tropics of plant protection by bringing-out leaflets, pamphlets, posters etc.

(iv) Scrutinizing and editing of scientific articles on plant protection for publishing Quarterly Journal "Plant Protection Bulletin".

(v) Maintenance/operation of library for consultation of technical journals/books,

(vi) Creating awareness among farmers on the plant protection technology by displaying exhibits in Melas, Shivirs etc.

Setup and Organisation

Deputy Director (PP)

Assist- Director (E)

Tech. Officer-Ill (2)

Artist Cum photographer

Artist Librarian

Centrally Sponsored scheme for Control of Pests and Diseases

Objectives:

To supplement state efforts by extending subsidies for controlling pests of agricultural importance.

Functions:

Extend financial assistance of the States/UTS for controlling selected pests in identified crops.

Activities:

(i) Assist DAC in:

(a) According administrative approval after scrutiny of proposals received from the States/UTs.

(b) Release of funds to the States/UTs.

(c) Settlement of accounts based on evaluation reports, final claims and audited figures of expenditure received from the States/UTs.

(ii) Convene Technical Committee meeting to consider proposals from States for including pests under the preview of the scheme.

Organisational Setup

Joint Director (E-II)

Technical officer -III

The Work done by Central Plant Protection

The important work done by this organization is given below:

Selection of Scheduled Desert Area of Locust: The scheduled desert area has been divided into three namely (i) central sub-division (ii) eastern sub-division and (iii) western sub-division. The head office of these sub-regions have been located at Jodhpur, Bikaner and Palampur, respectively. These sub divisions have again divided into zones and out posts. Usually each sub-division comprises of 8-12 zones and 20-90 out posts which has now been opened in the ratio of 10 : 40.

Locust Plague: Our country is suffering from the attack of locust, from very beginning but written proof is available from 1781. The locust swarms have been recorded periodically in the years noted below.

1812, 1821, 1834, 1843, 1863, 1864, 1878, 1889, 1896-97, 1901-03, 1906-07, 1912-15, 1926-31, 1940-46, 1948-63.

After the establishment of Directorate of Plant Protection in 1946, the locust plague could not become so harmful and wide spread. The locust swarms came after 1955 are summarised in the following table:

Year	*Swarns*	*Square miles breading area*
1959	24	6,000
1960	109	37,000
1961	79	50,000
1965		
1969 Could not become more active		
1974		

Looking to the damage caused by locusts the government sanctioned a co-ordinated anti locust scheme and use of helicopters and gliders in locust control. The helicopters and gliders were used during 1950-56 with the help of technical assistance programme of U.S.A. During 1955-56 the helicopters were used for the spray of insecticide to control the pyrilla in Madhya Pradesh. Looking to its successful use a permanent insect and plant disease control unit was established in air wing during 1957 which is being used time to time in most of the states.

Locust Investigatioins

Scheme was sanctioned in 1930 to survey locust breeding places and to study the life cycle of desert locust. But so far as the research unit is concerned it was started from 1954 as Locust Research Unit. The main function of this scheme is to gather as much information as possible related to locusts. Under this unit, some of the work was given to Forest Research Institute, Dehradun and some to Indian Agricultural Research Institute, New Delhi and rest to the Directorate of Bikaner. After some time Indian Agricultural Research Institute stopped to function and the work of Dehradun was handed over to Zoological Survey of India, Calcutta. To intensify the work, the field station for investigation on locusts was founded at Bikaner in 1975. In the establishment of this centre, Food Agriculture Organization has provided technical assistance.

Control of Flying Fox

In Northern India flying foxes caused considerable loss to fruits specially in Uttar Pradesh, Punjab and Himachal Pradesh. The Directorate of Central Plant Protection has conducted many experiments and found that they may be successfully controlled by the use of high frequency sound bomb.

National Rodent Control Campaign

The rats cause greater damage to food grains in comparison to other pests because they go on attacking and damaging grains right from the time it is sown till it is harvested and consumed. Therefore, a scheme on National Rodent Control was run under the guidance of Indian Council of Agriculture Research. In which the rodent control strategy has been divided in three phases. In the first phase (January-June) the operational training which consists of techniques of rodents control in crop fields, threshing floor and rural residential places, preparation of poison baits, their handling precautions to be taken and first-aid measures to be adopted in case of accidental poisoning should be imparted to the trainers student squads, village volunteers, young group etc. with the help of Agricultural Universities, Directorate of Agriculture and Pest Control Organisation etc.

In phase II (April-June) prior to the actual execution of the programme, the rural community will be prepared for rodent control through various extension agencies, *viz.* All India Radio, TV. and Film shows etc. In third phase actual execution of programme i.e. control operation will be followed in crops field (In Kharif-May and June- and in Rabi—September and October), threshing yard and residential houses (May, June).

Zincphosphide, warfarin, fomorin, barium carbonate, vacor, celphos and cymag may be used as rodenticides for their control. The dead rats may be collected and forwarded to Central Leather Research Institute, Madras for the economic utilisation of rat skins.

Central Plant Protection Stations

Due to the out break of some pests and diseases various states of the country become helpless to overcome the problem, and they demanded central assistance to solve the problem. Therefore,. fourteen central plant protection stations were opened at different places of the country. These are located at New Delhi, Pathankot Barabanki, Bilaspur, Amrawati, Cuttack, Dharwar, Hyderabad, Gaya, Indore, Tlisachirapalli, Ernakulu, Palampur and Guahati. The major work of these stations is to survey the insect pests, diseases, to find the methods of their control and to provide technical knowledge to their respective states. They also provide plant protection equipments and apparatus to the people on their demand. The Plant Protection officer is the incharge of each station.

Plant Protection Organisation in Uttar Pradesh and its Functions: Until 1946, there was no agency in the state to carry out results of research in plant protection at cultivators field and pests and diseases continued to take a heavy toll of crops and stored agricultural commodities running into crores of rupees. It was only in August 1947 that the State Government established plant protection service as a part of the drive to 'Grow More Food' with a view to help the cultivators in harvesting better and increased yield of crops by the adoption of timely plant protection measures.

Uttar Pradesh was the first state in the country to establish in Plant Protection Organisation in the year 1946-47. During pre Plan period (1947-48 to 1950-51) the total strength of plant protection organisation was 74 and centres and sub-centres were established in six regional districts and 11 other districts. The staff was located at the district headquarters only.

During the first plan period (1951-52 to 1955-56) the strength of the organisation increased to 230 and four new plant protection centres were established at Haldwani, District Nainital, Etawah, Dehradun and Mirzapur. During the second plan (1956-57 to 1960-61), plant protection centres/sub-centres were established in all the existing districts of the state. The strength of the staff increased to about 350.

During the third plan (1961-62 to 1965-66) the headquarter of this organisation was shifted from Kanpur to Lucknow with a post of Deputy Director, plant protection, as incharge of the scheme. During this period, 10 Mobile Training Units were created in each of this regions and each unit was provided with a senior plant protection/Assistant and a mobile van for imparting training to the farmers in simple methods of plant protection in all the districts of his region. During the three annual plans. (66-67, 67-68 and 68-69) plant protection sub-centres were established in the six hill districts, *i.e.*, Pithoragarh, Almora, Pauri-Garhwal, Chamoli, Tehri-Garhwal and Uttarkashi. Besides this, about 433 plant protection units at the block level as well as four campaign squads, one each at Meerut, Bareilly, Faizabad and Varanasi were established during this period. The total strength of the staff by the end of 1968-69 stood at 2141.

During the fourth plan period (1969-70 to 1973-74) all the Subject Matter Specialists (Plant protection) as well as Assistant plant protection officers were shifted to various district and gradually each of the district was manned by an officer. During this period all the remaining blocks were also manned with staff.

The general pattern of staff in the organisation at the moment is as follows:

At the headquarter, there is one Assistant Director Plant Protection, two Plant Protection Officers, six Senior Plant Protection Assistant, one Senior Mechanical Assistant and 8 members each of S.A.S. Group II and III supported by accounts and ministerial staff.

At the regional level, the plant protection staff is under the control of the regional Deputy Director of Agriculture who is assisted by a Senior Plant Protection Assistant.

At the district level, there is one Plant Protection Officer, one member each of S.A.S. Group I, II and III. However, in the regional districts there is one additional member each of S.A.S. Gr. I,II and III. In each of the 875 blocks in the state there is one A.D.O. (P.P.), plant protection supervisor and one field attendant. However, recently one additional field attendant has been provided in about 100 blocks where the receipt of the unit is above Rs. 25,000/-.

Plant Protection Organisation in U.P.

SI. No.	*Name of Post*	*State level*	*Region*	*District*	*Block*
1.	Joint Director (P.P.) Class—I.	1	—	—	—
2.	Deputy Director (P.P.) Class—I	—	1	—	—
3.	Assistant Director (P.P.) Class—II	1	—	—	—
4.	Assistant Director (P.P.) (Surveillance) Class —II for hill region	—	1	—	—
5.	Plant Protection Officer	1	—	1	—
6.	Account Officer	1	—	—	—
7.	Senior Plant Protection Assistant Group—I	7	—	1	—
8.	Assistant Development Officer (Plant Protection) Group—II.	8	—	1	1
9.	Plant Protection Supervisor Group—III	4	—	1	1
10.	Senior Mechanical Assistant Group—I	1	—	—	—
11.	Head Clerk	1	—	—	—
12.	Steno 1	1	—	—	
13.	Accountant	3	—	—	
14.	Assistant Accountant	2	—	—	
15.	Senior Clerk	12	1	1	—
16.	Junior Clerk	6	—	1	—
17.	Cashier	1	—	—	—

Contd...

Sl. No. Name of Post	State level	Region	District	Block
18. Mechanic	—	—	1	—
19. Field attendant/orderly Peon	6	1	5	1
20. Watchman	1	—	1	—
21. Driver 2	—	1	—	
22. Cleaner	1	—	—	—
23. Sweeper	1	—	1	—

Functions:

The main functions of the organisation are as follows:

1. To organize and maintain system of quick intelligence about agricultural pests and diseases attack in different parts of Uttar Pradesh so as to enable the farmers to take prompt and effective action against them.
2. To organize control operations against pests and diseases in the fields, orchards, kitchen gardens and godowns of farmers, fruit-growers and others on a wide scale.
3. To adopt measures for preventing the spread of agricultural pests and diseases within the state.
4. To demonstrate to the satisfaction of the public the effectiveness and suitability of recommended measures of control against diseases and weeds etc.; and to impress upon them the advantages of undertaking these measures.
5. To render advice and practical assistance to farmers, fruit growers and others to fight pests/diseases and weeds and thus raise more and better crops and save their crop produce from damage during storage.
6. To train extension workers and farmers in simple methods of plant protection including the use of plant protection machines.
7. To have an effective control on the quality of pesticides sold in the market.

4

Categories of Insects

Classification of Insects

There are about 15,00,000 insects which have already been recorded by Metcalf till 1940. These differ from each other in structure and habit. Each one of this kind is a distinct species, and is given a name by which it can be called. A group of species have some definite relationships known as a genus. Similarly, a group of allied genera form the subfamily, groups of subfamilies, the family and the families exhibit similar characters form an order. Several orders of similar identities form the class of Insecta phylum Arthropoda of the animal kingdom.

The foundation of the modern system of classification was laid by Brauer 1885. He considered the following points while classifying the insects:

(i) Presence or absence of wings.

(ii) Types and structure of mouthparts.

(iii) Metamorphosis.

(iv) Number of Malpighian tubules.

(v) Number of wings and their origin, segments of thorax and some other characteristics.

He divided class insecta into primitive wingless insects, the Apterygogenea, and the winged or secondarily wingless-insects, the Pterygogenea and included 17 orders under these two sub-classes. Borner (1904) recognized 24 orders of insects, while Handlirsch (1908) published a very different system on the fossil insects. He modified it in 1926 and included 34 orders. Imms (1961) reported 29 orders of insects which are widely accepted and are given as follows:

The class insecta may be divided into two sub-classes:

1. Apterygota, 2. Pterygota

Apterygota: The following are the main characters:

(i) These are the minute and primitive insects.

(ii) Wings are absent.

(iii) Development is generally direct and metamorphosis is absent.

(iv) One or more than a pair of pregenital appendages are present.

(v) Mandibles are situated on the anterior part of the head-capsule and move with the head capsule at a single point.

This sub-class includes the following orders:

1. Thysanura — Silverfish
2. Diplura — Japygids
3. Protura — Proturans
4. Collembola — Spring tails

Pterygota: Main characteristics of this sub-class are as follows:

(i) They are generally winged or secondarily wingless insects.

(ii) Metamorphosis is of different types and even some times wanting.

(iii) Pregenital appendages are absent.

(iv) Mandibles generally move with the head capsule at two points.

This sub-class may be sub-divided into two divisions:

(A) *Exopterygota:* The following are the main characters:

(i) Metamorphosis simple, sometimes slight, pupal instar rarely present.

(ii) Wings develop externally.

(iii) Immature stages are nymphs which resemble with adults in structure and habits.

It includes the following sixteen orders:

1. Ephemeroptera	Mayfly
2. Odonata	Dragonfly
3. Plecoptera	Stonefly
4. Grylloblattodea	Grylloblata
5. Orthoptera	Grasshopper
6. Phasmida	Phasmids
7. Dermaptera	Earwigs
8. Dictyoptera	Cockroaches, Mantids
9. Embioptera	Embids
10. Isoptera	Termites
11. Zoraptera	Zorapterans
12. Psocoptera	Book-lice
13. Mallophaga	Bird-lice
14. Siphunculata	Sucking-lice
15. Hemiptera	Plant bugs
16. Thysanoptera	Thrips

(B) *Endopterygota:* The following are the main characteristics of this division:

(i) Metamorphosis complete and accompanied by the pupal instar.

(ii) Immature stages are larvae which differ from adults in structure and habits.

(iii) Wings develop internally.

The following nine orders are included under this division :

1. Neuroptera	Lace Wings
2. Mecoptera	Scorpion fly
3. Lepidoptera	Moths and butterflies
4. Trichoptera	Caddis fly
5. Diptera	Houseflies and Mosquitoes
6. Siphonoptera	Fleas
7. Hymenoptera	Honeybees; wasps
8. Coleoptera	Beetles; Weevils
9. Strepsiptera	Stylops

The orders are of agricultural importance.

Order—Orthoptera

(Orthoptera, Greek orthos = straight, pteron = wing).

Characteristics:

1. Medium to large sized insects.
2. Mouthparts are typically biting and chewing type.
3. The antennae are generally elongated and filiform or setaceous type.
4. The majority of insects have a well developed prothorax and two pairs of wings, the fore wings being thickened and leathery in texture to form tegmina. The hind wings are membranous, larger than the front pair and folded fan like under fore wings.
5. In some groups the wings are small and half grown or lacking altogether, this condition is known as brachypterous form.
6. A pair of compound eyes and two or three ocelli are present.

7. Hind legs are modified for jumping, the coxae are small and widely separated, tarsi generally three or four cemented.
8. Cerci are present and usually short.
9. Metamorphosis is incomplete and the young ones look like an adult in structure and habit.
10. A thin cuticular membrane or tympanum is present on each side of the first abdominal segment or near the proximal end of each fore tibia.

Some Families of Agricultural Importance

(1) Family Acrididae

Characteristics:

1. The antennae are smaller than body length.
2. The prothorax extends down on the sides reaching the base of fore legs.
3. Hind legs are well developed with enlarged femora which are adapted for jumping. There is a row of peg like files on the femur used for sound production.
4. The tarsi are three segmented.
5. Tympanum is located on each side of the first abdominal segment.
6. The ovipositor is small and made of two pairs of valves.

Examples:

1. Rice grasshopper *(Hieroglyphus banian)*
2. Kharif grasshopper *(H. nigrorepletus)*
3. Surface grasshopper *(Chrotogonus spp.)*
4. Aak grasshopper *(Poecilocems pictus)*
5. Locust *(Schitocerca gregaria)*

(2) Family Tettigonidae (Locustidae)

Characteristics:

1. Long horned grasshoppers.

2. Apterous forms are common, when winged, the left tegmina usually overlaps the right one.
3. In males the cubito-anal regions of the tegmina are modified asymmetrically for stridulation.
4. In some groups the hind wings are absent and the tegmina reduced to the stridulating areas only.
5. The eggs are not enclosed in pods,.and the ovipositor frequently attains a great length, even exceeding that of body.

Example: Tettigonie viridissima.

Order—Isoptera

(Isoptera, Greek, isos = equal, pteron = wing)

Characteristics:

1. Minute to medium, soft bodied insects.
2. These are social insects of which different castes live together in colonies.
3. Winged members with two pairs of long narrow wings, which are similar in form, size and texture.
4. Mouth parts are of biting and chewing type.
5. Metamorphosis is simple.
6. The antennae smaller than the body length and of moniliform type.
7. Usually compound eyes and ocelli are absent, if ocelli are present, they are two in number.
8. The tarsi are four segmented.
9. Cerci is small and some time rudimentary.
10. The sexual organs are, with some exceptions, present in the reproductive castes only, the soldiers and workers showing various degrees of degeneration.

Some Families of Agricultural Importance

(1) Family—Mastotermitidae

Characteristics:

1. Tarsi are five segmented.

2. The hind wings having well developed anal lobe.
3. Thorax is broad.

Example: Mastotermes darwiniensis.

(2) *Family-Kalotermitidae*

Characteristics:

1. Worker caste is absent.
2. Ocelli are usually present.
3. Pronotum flat, usually broader than head.
4. Tarsi are four segmented.
5. Anterior wing scale long.

Examples:

(i) *Kalotennes incicus.*

(ii) *Cryptotermes domesticus.*

(3) *Family-Termitidae*

Characteristics:

1. The worker caste is well developed.
2. Pronotum of workers and soldiers are narrow with raised median anterior lobe.
3. The compound eyes are present.
4. The wings very slightly reticulate and anterior wing scale short.

Examples:

(i) *Odontotermes obesus.*

(ii) *Macrotermes serrulatus.*

(iii) *Eutermes heimi.*

(4) *Family-Hodotermitidae*

Characteristics:

1. Workers are present only in some genera.
2. Ocelli are absent.

3. Pronotum saddle shape, narrower than head.
4. Anterior wing scale short.

Examples:

(i) *Hodotermes macrocephalus*

(ii) *H. Konigi*

(5) Family- Rhinotermitdae

Characteristics:

1. The worker caste is present.
2. The pronotum of workers and soldiers is flat.
3. Anterior wing scales are usually long.
4. Compound eyes are present:

Examples:

(i) *Coptotermes ceylonicus*

(ii) *Rhinotermes sp.*

Order—Hemiptera

(Hemiptera Greek, Hemi = half, pteron = wing; means partly thickened and partly membranous wings.)

Characteristics:

1. Minute to medium sized, soft bodied, elongated oval insects.
2. Mouth parts are piercing and sucking type in which labium is adapted in the form of a beak or rostrum.
3. The compound eyes are generally large.
4. The antennae are four to ten segmented, the individual segments frequently long.
5. Two pairs of wings are present.
6. The cerci are absent.
7. Tarsi are one to three segmented.

The Hemiptera is redivided into two sub-orders:

(a) Sub order-Homoptera-Aphids, scale insects and leaf hoppers etc.

(b) Sub order: Heteroptera-Bugs.

Sub-order: Homoptera

Characteristics:

1. Generally two pairs of wings (except male scale insects) are present. The anterior pair is often of harder consistency and throughout of the same texture. Some times wings may be absent in some individuals.
2. Pronotum is small.
3. Tarsi are one to three segmented.
4. Metamorphosis is simple excepting male scale insects and white flies.
5. The distal end of labium is attached with the head.

Some Families of Agricultural Importance

(1) Family—Jassidae or Cicadellidae

Characteristics:

1. Generally small, elongate, cylindrical insects having eight segmented setaceous type antennae.
2. The double row of spines are present on the hind tibia.
3. The compound eyes are large situated on the lateral surfaces of the head and two ocelli are also present.
4. Thorax is narrower than head and attached with the abdomen without any prominent division.
5. Tarsi are three segmented.
6. Mostly rest on the lower surface of leaves and when disturbed they leap often several feet and fly.
7. The leaf hoppers are usually about 1/4 inch in length and usually tapering posteriorly.

Examples:

(i) Mango leaf hopper *Amritodus (Idiocerus) atkinsoni*

(ii) *Paddy leaf hopper (Nephotettix spp.)*

(iii) Cotton Jassid *(Amrasca bigutella)*

(2) *Family—Fulgoridae*

Characteristics:

1. The head in the form of a snout protruded forward.
2. The antennae are three segmented.
3. Generally two ocelli are present.
4. The wings are membranous, small and not fully covering the abdomen.
5. Tarsi five segmented.
6. They are mostly grey insects and the female is provided with anal tuft.

Example: Sugarcane leaf hopper *(Pyrilla perpusilla).*

(3) *Family-Aphididae*

Characteristics:

1. They are small delicate and more or less pear shaped insects.
2. Both winged and wingless organisms are found.
3. The head is hypognathous, small and fused with thorax.
4. The antennae are three to six segmented.
5. The compound eyes and three ocelli are present.
6. The tarsi are two segmented with paired claws.
7. The chronicles or honey-tubes are the most characteristic organs (absent only in few species) located on the 5th abdominal segment.
8. In the 9th tergum of abdomen, anus is situated which secrete honey dew.

Example: Mustard aphid *(Lipaphis erysimi)*

(4) *Family—Aleurodidae*

Characteristics:

1. The adults are small, four winged insects about three millimeters in length.

2. Their body is generally covered with a white waxy powder.
3. The vasiform orifice opens on the dorsal surface of the last abdominal segment. The opening is covered with an operculum under which there is a tongue like organ termed ligula.
4. The anus opens at the base of the ligula and secretes honeydew.
5. The compound eyes and two ocelli are present.
6. Tarsi are two segmented with paired claws in between them there is a spur or pad.
7. Metamorphosis holometabolous type which is an exception in this family. The eggs hatch into larvae, with the first moult they lose their legs antennae, following the additional moults, the insect enters pupal stage from which winged adults emerge.

Examples:

(i) Sugar cane white fly *(Aleurolobus barodensis)*

(ii) Cotton white fly *(Bemesia tabaci)*

(5) Family—Coccidae

Characteristics:

1. The insects are quite small, body is compressed and covered with whitish meal.
2. The female is wingless and round in structure.
3. The males are generally winged and the fore wings are only well developed.
4. The legs are highly reduced and tarsi one segmented.
5. The metamorphosis is complete in male and incomplete in female.
6. The mouth parts are absent in males hence they do not feed in the adult stage.

Example: Mango mealy bug *(Drosicha mangiferae)*

(6) Family—Lacciferidae

Characteristics:

1. The insects of this family are mainly confined to the tropics and sub-tropics.
2. They are extremely small insects with irregular globular body.
3. The legs and abdominal spiracles are absent.
4. The antennae are vestigial and three or four segmented.
5. Both winged and wingless males are found having four ocelli.

Example: Lac insect *(Laccifer lacca).*

Sub-order: Heteroptera

(Hetero = different, Pteron = wings)

Characteristics:

1. Two pairs of wings are always found in which fore wing is thickened at the base and membranous at apical portions. These are called hemelytra.
2. Mouth parts are piercing and sucking type, without palpi and arising from the front part of the underside of the head.
3. Body usually broad and flattened dorsoventrally.
4. The metamorphosis is incomplete.

(7) Family—Pentatomidae

Characteristics:

1. Medium to large sized bugs.
2. The head is small, triangular, having a longitudinal style on its centre.
3. A pair of compound eyes and two ocelli are present.
4. Scutellum is very large, reaching about one-half or more distance from posterior margin of pronotum to the end of folded wings.

5. Antennae are generally five segmented.
6. The tarsi are two or three segmented.

Example: Painted bug *(Bagrada cruciferarum).*

(8) *Family-Coreidae*

Characteristics:

1. Minute to medium sized insects.
2. Both the rostrum and antennae are four segmented.
3. A pair of compound eyes and two ocelli are present.
4. Tarsi are three segmented.
5. Metathorax possesses a pair of odour glands which are capable of producing disagreeable odour.

Example: Gundhi bug *(Leptocorisa varicornis)*

(9) *Family—Pyrrhocoridae*

Characteristics:

1. Medium to large sized insects.
2. Both antennae and rostrum are four segmented.
3. The ocelli are absent.
4. The tarsi are three segmented.
5. The prothorax is large and its distal ends are broader than the mesothorax.
6. Ceneus is absent on the hemelytron.

Example: Red cotton bug *(Dysdercus cingulatus)*

Order—Thysanoptera

The insects included in this order are commonly called as thrips.

Characteristics:

1. Small or minute slender-bodied insect.
2. The antennae are shorter having 6-10 segments.
3. Asymmetrical piercing and sucking type of mouth parts.

4. Maxillary and labial palpi present.
5. Prothorax well developed.
6. Tarsi one or two segmented each with a protrusive vesicle.
7. Wings when present very narrow with reduced venation and long marginal setae.
8. Cerci are absent.
9. Metamorphosis accompanied by one or two inactive pupal instars.

(10) Family -Thripidae

Characteristics:

1. Antennae are 6-8 segmented, usually with a 1 or 2 segmented spicula style; 3rd and 4th segments not conical, with sense-cones.
2. Fore tarsus never with claw like appendages at base of 2nd segment.
3. Ovipositor is saw like.
4. In female, the apex of abdomen is conical while it is rounded in male.

Example: *Thrips tabaci.*

Order—Coleoptera

(Coleoptera, coles = sheath, pteron = wings)

The order comprises of largest number of insects. More than 220,000 (about 4% of total insects) have been described. It includes beetles and weevils and most of them feed on plants and stored commodities.

Characteristics:

1. Medium to large sized insects.
2. The head is hypognathous or prognathous, flat, cylindrical or prolonged into a snout.
3. The compound eyes are large and usually ocelli are absent.

4. Two pairs of wings are present in which front pair greatly thickened and sclerotized called elytra, the hind wings membranous and protected by the front pair.
5. Mouth parts are chewing and biting.
6. Metamorphosis is complete.
7. Larvae of the beetles are known as grubs which are having three pairs of thoracic legs while the snout beetle (weevil) grubs are legless.

The order Coleoptera may be classified into three sub-orders. Adephaga, Arcehostemmate and Polyphaga out of these sub-orders, Poloyphaga; includes the families of agricultural importance.

Some Families of Agricultural Importance

(1) Family—Dermestidae

Characteristics:

1. These are small, oval, plump beetles.
2. The head is hypognathous and antennae are small, eleven segmented, clavate type.
3. The compound eyes large and one ocellus is also found.
4. The wings are simple and completely cover the abdomen.
5. The larvae are caraboid type and remain covered with dense hairs.

Example: Khapra beetle *(Trogoderma granarium)*

(2) Family—Bruchidae or Lariidae

Characteristics:

1. These are hard, small and stout beetles.
2. The antennae are eleven segmented and serrate to clavate type.
3. The head is prolonged into a short quadrate beak.
4. The wing covers are short, leaving the tip in the abdomen exposed.

5. The tarsi are five segmented.
6. The metamorphosis is hypermetabola, *i.e.*, in first instar larvae the legs, antennae and eyes are well developed but absent in IInd instar larvae.

Example: Pulse beetle *(Callosobruchus chinensis)*

(3) Family— Chrysomelidae

Characteristics:

1. These are small to moderate sized insects.
2. The antennae are large generally eleven segmented and filiform or clavate type.
3. Tarsi are four segmented in which the third segment is enlarged to form a kidney shaped pad.
4. The wings are simple and completely cover the abdomen.
5. The abdomen is short and five segmented.
6. Most of the grubs are stout with short legs and antennae. A number of them have spines and some of the leaf mining species are long and flat.

Example: Red pumpkin beetle *(Raphidopalpa foveicollis)*

(4) Family — Curculionidae

Characteristics:

1. The adults are specified by the prolongation of the head into a distinct snout.
2. The mouth parts are located at the tip of the snout.
3. The antennae are generally geniculate, club or clavate type and two segmented.
4. The body, elytra and legs are very hard.
5. The tarsi are five segmented.
6. Abdomen is five segmented in which first and second segments fused together.
7. The larvae are legless grubs, generally having dark head and white body.

Example: Rice weevil *(Sitophilus oryzae)*

(5) Family — Coccinellidae

Characteristics:

1. They are moderately small, round, convex, shining beetles.
2. The colour is generally red, yellow with black or blue spots.
3. The antennae are short and clavate.
4. The tarsi are four segmented in which the third segment is minute situated between the pad like second segment and the fourth segment bearing the claws.
5. The wings are developed and completely cover the abdomen.
6. Some species are predaceous on aphids and scale insects while others are plant feeders.

Examples:

(i) Spotted beetle (*Epilachna vigintioctopunctata)*

(ii) Coccinellid beetle (*Coccinella septempunctata)*

(6) Family—Tenebrionidae

Characteristics:

1. The body is hard, often flattened, globular beetle.
2. The antennae are short under the projection of the sides of head.
3. The elytra fit closely to abdomen.
4. Some species are apterous with vestigial or no wings.
5. The first three visible abdominal segments are connate.
6. The larvae are elongate and cylindrical in which the hind end bears two dorsal hooks and a ventral retractile process.

Example: Red rust beetle (*Tribolium caslaneum)*

(7) Family—Bostrichidae

Characteristics:

1. The integument is hard and the pronotum is largely developed and hood like.

2. The body is truncated posteriorly and armed with small projections.
3. Trochanter distally obliquely joined to femora.
4. The antennae are club shaped and having less than two segments.
5. The larvae have a curved body enlarged at the thorax with developed legs, annular, spiracles and mandibles.

Example:(i) Red flour beetle *(Rhizopertha dominica)*

(8) Family- Scarabidae **(Melolonthidae)**

Characters:

1. The beetles are variously coloured, black, brown, green, mahogany or blue.
2. The mentum and ligula are fused.
3. Beetles are commonly attracted to light.
4. Antennae are 8-11 segmented.
5. The elytra are short and do not cover the complete abdomen.
6. The grubs are fleshy, white with brown head, curved body (C- shaped larvae) and the larval segment large and smooth.

Examples:

(i) Groundnut whitegrub, Holotrichia consanguinea
(ii) *H. serrata*
(iii) Kurmula species —*Anomala spp.*
(iv) Melolonthid beetle—*Melolontha melolontha*

Order—Lepidoptera

(Greek Lepidoptera, Lepidos = scales, pteron = wings)

Characteristics:

1. The adults vary greatly in size from minute to large species and body is covered with scales.

2. They are the most familiar insects on account of their beautiful colouration, particularly among butterflies, and their abundance in vegetables and gardens.
3. The head is small, hypognathous type and attached with the thorax by a cervix.
4. Compound eyes are comparatively large and the ocelli are generally present.
5. The mouth parts are greatly reduced and represented by a suctorial proboscis formed by the maxillae.
6. Generally, two pairs of wings are present, except in four species which have wingless females covered with scales.
7. The tarsi are five segmented.
8. The metamorphosis is complete and the larvae are generally called as caterpillars which are having chewing and biting type mouth parts.

This order is generally divided into Rhopalocera (butterflies) and Heterocera (moths) simply because of the difference in general appearance of the two groups. Some have divided the order into Macro and Micro Lepidoptera simply on the basis of size. According to Imms (1961) the order may be divided into following three suborders. His classification was based on work of several authorities and hence widely accepted ones.

1. Zeugloptera
2. Monotrysia
3. Ditrysia

Suborder—Zeugloptera

1. Adults with functional mandibles.
2. Lacinia is well developed and galea is not forming a proboscis.
3. The female genital chamber opens in the main oviduct.

Suborder—Monotrysia

1. Mandibles are vestigial.

2. Lacinia is absent.
3. Females with one or two genital openings behind the 9th abdominal sternite.
4. The wings are generally aculeate.

Suborder—Ditrysia

This is the largest suborder and includes all the families of agricultural importance. The distinguished features are as follows:

1. The female has a copulatory pore on the 8th sternite and an egg pod on the 9th sternum.
2. The fore-wings are without jugam lobe.
3. The hind wings are often with frenulum and reduced venation.

Some Families of Agricultural Importance

(1) Family — Gelechiidae

Characteristics:

1. These are small to medium sized moths of dull colour.
2. The fore wings are narrow than the hind wings.
3. Maxillary palpi are either absent or rudimentary.
4. The labial palpi are three segmented and directed anteriorly.
5. The hind tibia possess hard hairs.

Examples:

(i) Cotton pink boll worm *(Pectinophora gossypiella).*
(ii) Potato tuber moth *(Gnorimoschema operculella).*
(iii) Angoumois grain moth *(Sitotroga cerealella)*

(2) Family — Pyralididae

Characteristics:

1. These are small or moderate size moths with the labial palpi projecting snout like.

2. Compound eyes and ocelli are present.
3. The four wings are narrow while hindwings are broad.
4. The larvae are recognized by the presence of two setae and by the arrangement of the characters of the prolegs in a circle.
5. The larvae feed on leaves, fruits, flowers and stems and most of them are borers.

Examples:

(i) Cotton leaf roller *(Sylcpta derogata).*

(ii) Juar stem borer (*Chilo partellus).*

(iii) Sugarcane top borer *(Tryporyza nivella).*

(iv) Sugarcane root borer *(Emmalocera depresella).*

(v) Sugarcane stem borer *(Acigona stenillus).*

(vi) Sugarcane stem borer *(Chilotraea infuscatellus).*

(3) Family — Noctuidae

Characteristics:

1. Medium to large sized insects and generally of brown colour.
2. These are nocturnal in habit and light lover.
3. The maxillary palpi are absent while labial palpi are present.
4. The antennae are filiform in females and pectinate in males.
5. The fore wings are rather narrow and some what elongate.
6. The larvae are of dull colour, and generally provided with the full complement of prolegs.

Examples:

(i) Gram cut worm *(Agrotis psyilon).*

(ii) Gram caterpillar *(Helicoverpa armigera).*

(iii) Fruit sucking moth *(Ophideres conjuncta).*

(iv) Fruit sucking moth *(Achoea janata)*.

(v) Linseed semilooper *(Plusia sp.)*.

(4) *Family — Cymbidae*

Characteristics:

1. Minute to medium sized insects.
2. They are frequently green insects found among the herbage of trees and shrubs.
3. The wings are broad, spotted or marked with streaks.
4. The larvae are never prominently hairy and the cocoon is boat shaped.

Example: Spotted boll worm *(Earias vitella E. insulana E. cumateria)*

(5) *Family — Arctiidae*

Characteristics:

1. These are usually stout bodied moths, often with moderately broad wings, which are conspicuously spotted, banded or otherwise marked with bright colours.
2. Generally nocturnal in habit and are attracted to light.
3. The larvae are clothed with dense long hairs.

Example: Red hairy caterpillar *(Amsacta moorei)*

(6) *Family—Saturnidae*

Characteristics:

1. These are stout hairy and broad winged insects.
2. Antennae are bipectinate in both the sexes.
3. The labial palpi are minute and there is no frenulum.
4. The head is more or less shrunken into the prothorax.
5. The caterpillars are stout bodied with spiny tubercles or tufts of hairs and may attain the length of 7 to 8 cm.

8. Caterpillars feed on leaves and spin large cocoons.

Examples:

(i) Eri silk worm *(Attacus ricini).*

(ii) Tasar silk worm *(Antherea paphia).*

(iii) Munga silk worm *(A. assama).*

(iv) Wild silk moth *(Attacus atlas).*

(7) Family— Bombycidae

Characteristics:

1. These are medium sized and yellowish white insects.
2. The antennae are pectinate in both the sexes.
3. The larvae are glabrous and elongate having anal horn on the 8th abdominal segment.
4. The larvae form dense silken cocoons.

Example: Silk worm *(Bombyx mori).*

(8) Family-Pieridae

Characteristics:

1. These are generally medium sized butterflies.
2. The colour of wings is almost yellow, orange or white, usually marked with black.
3. The claws of legs are bifid and toothed.
4. The larvae are green and covered with fine and short hairs.
5. There is a cephalic spine on the head of the pupa.

Example: White butterfly *(Pieris brassicae).*

(9) Family— Papillionidae

Characteristics:

1. These are generally large butterflies of bright colour.
2. The wings are black and are generally marked with yellow and frequently with green and blue.
3. The hind wings possess tail like prolongation.

4. The caterpillars are naked or sometimes covered with fine hairs,
5. They are always provided with an organ known as osmeterium.

Example: Citrus butterfly *(Papileo demoleus).*

Difference between Moth and Butterfly

Moth	*Butterfly*
1. They are generally yellowish white in colour.	1. These are bright coloured.
2. Nocturnal in habit.	2. Diurnal in habit.
3. The antennae of moths are variable in forms but are generally filamentous.	3. The antennae are generally clavate or capitate type.
4.. Generally two ocelli are present.	4. The occelli are absent.
5. The wings at rest are held horizontal position.	5. The wings are held in vertically at rest.
6. The pupae are generally covered with cocoon.	6. The pupae are naked known as chrysalis.
7. The eggs are generally flat and round.	7. Eggs are cigar shaped and cylindrical.
8. The larvae are generally covered with hairs.	8. The larvae are smooth and naked.

Order—Hymenoptera

(Hymenoptera, Hymen = membrane, pteron = wings).

Characteristics:

1. Generally two pairs of membranous wings are present and hind pair being smaller.
2. Small to minute sized insects.
3. The venation on wings is more or less reduced.
4. The compound eyes are well developed and generally three ocelli are present.

5. The antennae range from 3 to 60 segmented and are filiform, pectinate or geniculate type.
6. Abdomen of female is generally provided with a saw, piercing organ or sting.
7. The tarsi are five segmented and provided with a pair of claws and empodium.
8. Metamorphosis is complete.
9. The larvae are caterpillar or grub like, and having a distinct head and chewing mouth parts, some with thoracic or abdominal legs or both, and others without any legs.

Some Families of Agricultural Importance

(1) Family—Apidae

Characteristics:

1. It includes a vast range of solitary and parasitic bees and three principal groups of social species.
2. The head is as broad as thorax.
3. Each corner of the pronotum ends in a small round lobe.
4. The body hairs are branched and give a velvety appearance.
5. The maxillary palpi are one segmented and labial palps four segmented.
6. The hind basitarsus are widened and provided with long hairs.
7. A colony of bees is composed of three castes, a queen, workers and males (drones).

Examples:

(i) *Apis indica*

(ii) *Apis dorsata*

(iii) *Apis florea*

(2) *Family—Tenthredinidae*

Characteristics:

1. These are small to medium sized saw flies.
2. The posterior region of scutellum is separated as postscutellum.
3. The ovipositor consists of plate.
4. The claws are toothed.
5. The larvae are caterpillar like in form and bear thoracic legs and also usually abdominal prolegs.

Examples: Mustard sawfly *(Ahalia proximo)*

(3) *Family—Braconidae*

Characteristics:

1. These are the minute to small sized insects and are parasitic ones.
2. The compound eyes are naked and three ocelli are present.
3. The hind femur is simple and thickened.
4. The ovipositor is about the length of body.
5. Polyembryony is known to occur in several species.

Examples:

(i) *Bracon sp.*

(ii) *Chelonus sp.*

(iii) *Apanteles sp.*

(iv) *Microgaster sp.*

(4) *Family—Chalcididae*

Characteristics:

1. These are small having reduced wing vernation and parasitic insects.
2. The antennae are generally geniculate type and small.
3. The ovipositor is small and erect.

4. The chalcid flies generally attack the egg and larval stages of their hosts.
5. The thorax is comparatively large and provided with small legs.

Examples:

(*i*) *Trichogramma minutum*

(*ii*) *Chalcis sp.*

(*iii*) *Brachymeria sp.*

(5) *Family—Formicidae*

Characteristics:

1. These are social insects and colony consists of three castes, workers, males and females.
2. One or two nodes are present between the propodeum and gaster (swollen part of the abdomen).
3. The antennae are geniculate type.
4. The wings are generally absent except in the sexually mature forms. After mating the males die and the wings of young female are shed.
5. The joints of head, throat and abdomen are very clear.
6. The labrum is vestigeal, maxillae one to six segmented, the lacinia are membranous and without teeth.
7. The food habits of ants differ very much.

Example: Ants.

(6) *Family —Dchneumonidae*

Characteristics:

1. The insects are parasites or less frequently hyperparasites of Lepidoptera.
2. Costal cell narrow or obliterated.
3. Sternites of gaster eventysclerotized.
4. Fore-wing with cross vein 2m-cu is generally present.

5. Hind wing with cross-vein r-m meeting SC + R before Rs leaves it.

Example: *Thersilochus conotracheli*

Order—Diptera

(Diptera, Di = twice; pteron = wings).

Characteristics:

1. Minute to medium sized insects commonly known as flies and mosquitoes.
2. The meso thorax bears one pair of wings and the second pair is represented by halteres.
3. The mouth parts are adapted for piercing and sucking or for lapping or sponging. The mandibles are generally absent.
4. The pro and meso-thorax are comparatively small while the meta thorax is large.
5. The tarsi are five segmented.
6. Metamorphosis is complete.
7. The larvae generally called as maggots which are legless and most of them have no distinct head.

Some Families of Agricultural Importance

(1) Family—Trypetidae or Tephritidae

Characteristics:

1. These are small to medium sized flies.
2. The wings are generally banded or spotted.
3. The flies are yellowish or brownish in colour.
4. Ovipositor is well developed and three segmented.
5. The antennae are three segmented and aristate type.
6. The apical part of sub-costa bends abruptly forward and then become weak.

Example: Fruit fly (Strumeta) *Dactis cucurbitae.*

(2) *Family-Tachinidae*

Characteristics:

1. These are small and parasitic insects.
2. The head is large and free.
3. The post-scutellum is not fully developed.
4. The antennal hair, arista is generally naked.
5. The members of this family attack larvae or adults of many orders specially lepidopterous larvae.

Example: *Strumia cericariae* (Parasite of the larvae of silk worm)

(3) *Family—Syrphidae*

Characteristics:

1. These are medium to large size flies and bright coloured.
2. The head is large.
3. The antennae are three segmented and arise from the short or large tubercles.
4. The flies fly for a pretty long time on shrubs or herbs.

Example: *Syrphus sp.*

(4) *Family—Cecidomyidae*

Characteristics:

1. Minute delicate flies with long moniliform antennae.
2. Ocelli present or absent.
3. Wings are with longitudinal veins.
4. Coxae are not elongate and tibial avoid of spurs.

Examples: Linseed gallmidge: *Dasyneura lini*

Paddy gall midge: *Patchydeplosis oryzae*

(5) *Family—Agromyzidae*

Characteristics:

1. The insects are small to minute flies.

2. The larvae are usually leaf miners.
3. Usually winged species, tarsi are quite different.
4. Posterior basitarsus normal.
5. Fore femur is without spine.
6. Post vertical bristles are divergent.

Example: Pea leaf miner (*Phytomyza atricornis*).

5

INSECT CONTROL

The Entomology consist from "Greek" words. The means of "Entomo" is insect and means of "Logos" is knowledge.

- Entomology is the branch of zoology that deals with the study of insects under class insecta of phylum "Arthropods".
- The lac insect *Laccifer lacca* Kerr belongs to the order Hemiptera.
- The silk worms belong to the insect order Lepidoptera.
- Honeybees belong to the order of "Hymenoptera".
- It has been estimated that 50-70% of the grain crops are pollinated by insects, a major portion of it by bees alone.
- The latest production of honey in India is estimated at about 25000-30000 tonnes, its major portion obtained from "Apis *dorsata*".
- The all insects deficit RBC in blood.
- The blood colour of insects is green with yellow.

- The latest estimated total annual losses caused by insect pests to major crops and the food-grains in storage in India to the Rs. 336.6 billion.
- The world wide pre-harvest and post-harvest food losses due to pests have been estimated to be about 48%.
- There is a higher percentage of 46% loss for rice and a lower percentage of 24% for wheat.
- The higher loses occur in Africa 42% and Asia 43% respectively of their potential crop value.
- By the controlling insect-pests of sugarcane the yield could be increased by 20% from 229.2 to 284.9 million tonnes leading to an additional income of 43730 millions of rupees.
- Plant genetic system, a Belgium Biotechnology company in July 1987 were the first reported the development of transgenic plants of tobacco.
- Over 2.5 billion kg of pesticides are applied in the world annually.

Pesticides Consumption durint 1999-2000 in India

Insecticides	60%
Fungicides	21%
Herbicides	14%
Others	5%

- Punjab have the major consumption technical grade pesticide in India, approximately 7150 metric tonnes.
- Lowest consumption of pesticide in India is Daman and Diu, approximately 1 metric tonnes.
- Bartlett (1956) coined the term "Integrated Pest Control"

The term "pest management" was given by Geier (1970).

Perfect Organisation

IPM as a dynamic and constantly evolving approach to crop protection in which all the suitable management tactics and available surveillance and forecasting information are utilized to develop a holistic management programme as part of a sustainable crop production technology (By Dhaliwal and Arora 2001)

- Destructive Insect and Pest Act was enforced in the year of 1971.
- The maximum percentage of insecticide is used in cotton in India.
- Average damage of food grains during stored period in India is 10%.
- Plant Protection and Quarantine Act was passed in the year of 1914.
- First insecticide discovered was D.D.T.
- The maximum amount of sugar in honey is 73%.
- 68% resin found in lac insect. *(U.P. Lecturship 2003).*
- Central Honey Bee Research Station is situated at Pune (Maharashtra).
- Lac Research Station is situated at (Bihar).
- Central Silk Research Station is situated at West Bengal.
- D.D.T. was first of all made by "Zedler" scientist.
- The first synthetic pyrethroid was described in 1973.
- LD_{50}: Lethal dose required for killing 50% population of test animals is known as LD_{50} value. The unit of LD_{50} is mg/kg of body weight of test animals or Hg/g body weight of insect.
- Biology has two subdivisions. BOTANY-which deals with plants and ZOOLOGY-which deals with animals.
- The aim of zoology is to collect knowledge of animals.

Zoology Branches

Entomology	—	Study of Insects
Protozoology	—	Study of Protozoa
Nematology	—	Study of Nematodes
Rodentology	—	Study of Rodents
Ornithology	—	Study of Birds
Mammology	—	Study of Mammals
Ichthyology	—	Study of Fishes
Helminthology	—	Study of Worms
Acarology	—	Study of Mites

Entomology Branches

Insect Taxonomy
Insect Morphology
Insect Anatomy
Insect Histology
Insect Physiology
Insect Toxicology
Insect Pathology Insect Parasitology
Insect Genetics

Applied Entomology

Agricultural Entomology
Medical Entomology
Horticultural Entomology
Veterinary Entomology
Storage Entomology
Forest Entomology
Industrial Entomology

Insect Structures

The Rasping-sucking mouth parts are intermediate in structure between the piercing-sucking and chewing type and are found in Thrips.

The Chewing-lapping type of mouthparts are found in honeybees.

The Piercing-sucking type of mouthparts are found in bugs and mosquitoes.

The Sponging-type of mouth parts are common in flies the typical example being that of the housefly.

- *Pest:* A pest is an organism whose population often increase above a certain level of economic injury and its existance conflicts with man's welfare, convenience and profits.

Sorts of Pests: There are three categories of pests

(i) *Regular Pests:* These are generally found in abundance during a crop season e.g. aphids, Jassids, and Thrips.

(ii) *Sporadic Pests:* These assume pest status occasionally in certain years and include locusts, grasshoppers, hairy caterpillars, crickets and cutworms.

(iii) *Potential Pests:* These pests normally cause negligible damage but may become highly destructive resulting from some disturbance in the environment and the consequent increase in their number e.g., armyworm on wheat.

Controlling Measures

(a) *Legislative Measures:* Legislative control involves enactment of laws to regulate the entry, establishment and spread of pests.

Legislation is also imperative to stop the accidental entry, from outside the country of certain pests which may not be present in that country.

Thus legislation is of four kinds

(i) Legislation for foreign quarantine to prevent the introduction of new pests from abroad.

(ii) Legislation for domestic quarantine to prevent the spread of established pests within the country or within a particular state.

(iii) Legislation for notified campaign of control against pests.

(iv) Legislation to prevent the adulteration and mishandling of insecticides or other devices used for the control of pests.

(b) *Cultural Practices:* The cultural methods of insect control comprise regular form operations, which are so performed as to destroy the insects or to prevent them from causing injury.

For achieving cultural control, the various agricultural practices can be categorised under various groups

1. Tillage
2. Clean seed
3. Irrigation
4. Fertilizers
5. Clean culture
6. Crop rotation
7. Trap crops
8. Pruning and Thining
9. Intercropping
10. Destruction of crop residues.

(c) *Mechanical and Physical Control:* Mechanical and physical control measures involve the use of force or physical factors of the environment with or without the aid of special equipment.

Labour's Part

It means working with hands. Sometimes with the aid of some simple equipment, like bags, nets etc.

(d) *Chemical Control:* The use of chemicals for the control of insect-pests. The classification of insecticides is as follows.

Division of Insecticides: Insecticides can be grouped in various ways:

According to the Mode of their Entry

(1) *Stomach Poisons:* They are generally used against insects with chewing type of mouthparts and under certain conditions against those with sponging, siphoning, lapping or sucking mouth parts.

(2) *Systemic Poisons:* A systemic insecticides when applied to seeds, roots, stems or leaves of plants is absorbed and translocated to various parts of the plant in amounts lethal to insects which feed on them, like-dimethoate, phosphamidon, phorate, aldicarb etc.

(3) *Contact Poisons:* The contact poisons are applied as sprays or dusts either directly onto the body of insects or to the places frequented by them. These poisons kill the insects either by clogging spiracles and respiratory system or by entering through the cuticle into the blood and acting as nerve or general tissue poisons. like-chlorinated hydrocarbons (DDT, HCH, aldrin etc.) carbamates (carbaryl, aldicarb, carbofuran etc.) and organophosphates (Malathion, Monocrotophos etc.)

(4) *Fumigants:* Poisonous gases, derived from either solids or liquids are used as fumigants to kill insect pests of stored grains and other products in warehouses, museum, godowns etc.

- The common fumigants are-sulphurdioxide, phosphine, methylbromide, EDB etc.

According to Mode of Action

Physiological poisons

Protoplasmic poisons

Respiratory poisons

Nerve Poisons

According to their Chemical Composition

(i) *Elements:* Such as sulphur, phosphorus, thallium, mercury etc.

(ii) *Inorganic Compounds:* Such as lead arsenate, sodium fluosilicate, zinc phosphide, paris green etc.

(iii) Organic compounds

(a) Compounds of plant origin such as pyrethrum, nicotine, rotenone etc.

(b) Animal and mineral oils such as fish oil, diesel oil etc.

(c) Synthetic organic compounds such as DDT, Malathion, carbaryl etc.

(iv) *Poisonous Gases:* Such as hydrogen cyanide ethylene dichloride, carbon tetrachloride, methyl bromide, phosphine etc. used as fumigants.

(a) *Nematicides:* These chemicals are used for controlling plant parasitic nematodes. Some of common nematicides are mentioned here.

(1) DBCP (2) D-D Mixture

(3) Ethylene dibromide (4) Ethoprop

(5) Phorate (6) Aldicarb

(b) *Rodenticides:* These chemicals are used for control of rodents such as - strychnine, warfarin, zinc phosphide etc.

(c) *Molluscicides:* Molluscisides may be classified into two broad categories.

(i) *Aquatic Molluscicides:* copper sulphate niclosamide, sodium pentachlorophenate etc.

(ii) *Terrestrial Molluscicides:* include carbamates-aminocarb, methiocarb, mexacarbate etc.

(d) *Fungicides:* These chemicals used against fungus like

(i) Bordeaux mixture

(ii) Sulphur and allied compounds

(iii) Carbamates

(iv) Organic mercury compounds

(v) Antibiotics

(vi) Nitrogen compounds Pesticide Application Equipment

The various equipment used include dusters, sprayers, vapour and smoke generators, agricultural aircrafts, granule applicators etc.

The Insecticide Dusters: The insecticides dust which are very fine particles when falling free in air either slowly or driff for a long distance due to wind.

- Parts of the Dusters

(i) *Hopper or Container:* It is used for holding the dust.

(ii) *Blower:* It work to throw the dust outside from container to the discharge outlet.

(iii) *Handle:* It is used to rotate the gear and there by fan throws the air from blower.

(iv) *Feeding Brush:* It agitates the dust in the hopper to prevent uneven dusting.

(v) *Feed Mechanism:* It control the rate of powder coming from hoppers.

(vi) *Discharge Line:* It carries the powder from bowler to nozzle.

(vii) *Shoulder Streps:* Shoulder streps are provided with mounting frame in some dusters to set whole mechanism on the chest.

Sorts of Dusters

(a) Manually Operated Dusters

  (i) Plunger duster

  (ii) Bellows dusters

  (iii) Rotary or fan dusters

  (iv) Wet dusting equipment

(b) Power Operated Dusters

  (i) Tractor mounted duster

  (ii) Engine-operated duster

Merits of Dusters

1. Large area can be covered
2. Less expensive
3. Easy to operate
4. Very effective in arid zone area Demerits

7. Increase toxic hazzard

8. Sprayers

A sprayer is an appliance which atomizes the spray fluid, which may be a suspension, an emulsion or a solution, the fluid is ejected with some force for proper distribution.

The various types of sprayers can be grouped into four categories, depending upon the volume of spray fluid discharge to cover a unit area.

1. *High Volume Sprayer:* These require 300-500 litres per ha. of spray fluid. For example-knapsack sprayer, stirrup sprayer, rocker sprayer, compression sprayer, foot sprayer etc.
2. *Low Volume Sprayer:* These sprayers require 50-100 litres of spray fluid per ha. for instance-motorized knapsack sprayers.
3. *Ultra Low Volume:* These sprayers require 1-5 litres per ha. of spray fluid. For example knapsack mist blowen fogair.
4. *Aerosols:* Less than 1.0 litres per ha. of spray fluid is required to create fog. The example are-pressurised containers, swing fog machine.

Kinds of Sprayers

(a) Manually Operated Sprayers

These are of two types

(i) Pneumatic sprayers

(a) Hand compression sprayers

(b) Pressure retaining pumps.

(c) Small pneumatic atomizers.

(ii) Hydraulic sprayers

(a) Knapsack sprayer (b) Foot sprayer

(c) Rocker sprayer (d) Stirrup sprayer

(b) Power Operated Sprayers

(i) Mist blower or motorized knapsack sprayer

(ii) Tractor mounted sprayer

(iii) Power sprayer

(iv) Ultra-low-volume sprayer

(v) Aerosol dispensers

(vi) Smoke generator

(vii) Vapour generators.

(c) Agricultural Aircrafts

(i) Light aircraft

(ii) Medium aircraft

(iii) Heavy aircraft

(d) Granule Applicator

Placement of insecticidal granules in the whorls of crops like-maize, jawar, etc., is very effective for the control of stem borers.

(e) Soil Injectors

Most soil fumigants are liquids and are applied with soil injectors.

- Merits of sprayers

1. Less drift problems as compared to dusters.
2. It is handy and simple to use.
3. Used to apply various types of chemical such as bactricide, fungicide, insecticide etc.
4. Most commonly used appliances.
5. Covers more area in less time.
6. Treatment can be done on height and also blow the leaves.

Sustainable pest management is otherwise known as IPM i.e., Integration of tactics of control a single pest on one or more crops. IPM as applied in agriculture is ideally the use of most effective, economical, safest, ecologically sustainable and sociologically acceptable combination of physical, chemical and biological methods to limit the harmful effects of crop pests.

According to FAO panel of experts (1966-72) IPM is-"A pest management system that in the context of the associated environment and the population dynamics of the pest species,

utilizes all suitable a manner as possible and maintains the pest population at levels below those causing economic injury."

The overall objective of IPM is to create and to maintain situations in which insects are prevented from causing significant damage to crops.

A number of tools or components have been successfully used for IPM in several crops. These include-use of pest resistant or tolerant, predators and pathogens, use of parasites, summer ploughing, quarantine measures, late planting, hard collection and destruction, judicious use of pesticides, attractants, repellents, sterilants, growth regulators.

- The sustainable pest management in India has yet to come to be well recognized, but it is still more of an aspiration than a reality for the average farmers.

Importance of Biopesticides: Biopesticides are most important, effective and commercially viable as these are inexpensive cause no pollution. Pose no risk to human health and have a long term capability to control pests. Biopesticide agents are abundantly available in nature. Several pathogens including viruses such as-NPV (Nuclear polyhedrosis viruses and granulosiss viruses (G.V.)

- Bacteria like - *Bacillus thuringiensis.*
- Fungi like-Metarhizium, Beauveria, Verticillium.
- Protozoa like- *Schizogregrine* cause diseases in insects and destroy them. Similarly, several insect-parasitoids (Parasites thriving on insects are also known in nature.
- *Trichoderma* is an egg parasitoid of several pests.

	Common Name	Scientific Name	Order	Host Plants	Damaging Stage	Mouth Parts
1.	Locust	*Schistocera gregaria* Forsk	Orthoptera	Polyphagous adult	Nymph and cutting type	Chewing and
2.	Termite	*Odontotermes obesus* Ramb	Isoptera	Polyphagous	Nymph and adult	Chewing and cutting type
3.	White Grub	*Holotrichia consanguinea*	Coleoptera	Polyphagous	Grub and adult	Biting and cutting type
4.	White Fly	*Alerolobus baradensis* Mask	Hemiptera Homoptera	Sugarcane, Jawar, Bajra, Wheat, Barley	Nymph and adult	Piercing and sucking type
5.	Leaf Hopper or Pyrilla	*Pyrilla perpusilla* Walk	Hemiptera Homoptera	-do-	-do-	-do-
6.	Sugarcane top borer	*Tryporyza nivella*	Lepidoptera	Sugarcane, Jawar, Bajra, Kans, Sarkanda	Larva	Biting and cutting type
7.	Shoot borer	*Chilo infuscatellus* Snell	Lipidoptera Jawar, Maize	Sugarcane,	Larva	Biting and cutting type

Contd....

	Common Name	*Scientific Name*	*Order*	*Host Plants*	*Damaging Stage*	*Mouth Parts*
8.	Paddy stem borer	*Tryporyza incertulas*	-do-	Paddy	-do-	-do-
9.	Rice gundhi bug	*leptocorisa varicornis*	Hemiptera	Paddy, Jawar, Bajra Maize	Nymph and adult	Piercing and sucking type
10.	Paddy plant hopper	*Nephotettix* spp.	Hemiptera	Paddy	-do-	-do-
11.	Cotton Jassid	*Amrasca bigutulla bigutulla*	Hemiptera	Cotton, bhindi	Nymph and adult	-do-
12.	Cotton leaf roller	*Sylepta derogata*	Lepidoptera	-do-	Larva	Chewing and cutting type
13.	Spotted boll worm	*Earias vitella*	-do-	-do-	-do-	-do-
14.	Pink boll worm	*Pectinophora gossypiella* Sound	Lepidoptera	-do-	-do-	-do-
15.	Mustard aphid	*Lipaphis erysimi*	Hemiptera	Radish, Rai, Mustard, Cauliflower, turnip etc.	Nymph and adult	Piercing and sucking type

Contd....

Common Name	*Scientific Name*	*Order*	*Host Plants*	*Damaging Stage*	*Mouth Parts*
16. Painted bug	*Bagrada cruciferorum*	Hemiptera	-do-	-do-	-do-
17. Cut worm	*Agrotis ypsilon*	Lepidoptera	Mainly gram and potato	Larva	Cutting and biting type
18. Gram caterpillar	*Helicoverpa armigera*	-do-	Gram, pea, tomato, Arhar	-do-	-do-
19. Red pumpkin beetle	*Raphidopalpa foveicollis*	Coleoptera	Plants of cucurbitaceae	Grub and adult	-do-
20. Fruit fly	*Dacus cucurbitae*	Diptera	Plants of cucurbitaceae	Maggot	-do-
21. Epilachna beetle	*Epilachna vigintioctapun-ctata*	Coleoptera	Brinjal and cucurbits	Grub and adult	-do-
22. Brinjal shoot and fruit borer	*Leucinodes arbonatis*	Lepidoptera	Brinjal and Potato	Larva	-do-
23. Diamond back moth	*Plutella maculipensis*	Lepidoptera	Plants of cruciferae	-do-	Cutting and biting type

Contd....

	Common Name	Scientific Name	Order	Host Plants	Damaging Stage	Mouth Parts
24.	Tobacco caterpillar	*Prodenia litura*	Lepidoptera	Tobacco and cruciferae	-do-	-do-
25.	Mango hopper	leaf *Amritkodus atkinsoni*	Hemiptera Homoptera	Mango adult	Nymph and sucking type	Piercing and
26.	Mango mealy bug	*Drosicha mangifera green*	Hemiptera Homoptera	Mango, Jack fruit etc.	-do-	-do-
27.	Lemon buterfl	*Popilio demoleus* Linn	Lepidoptera	Citrus	Larva	Biting and cutting type
28.	Ber fruit fly	*Carpomyia vesuviana*	Diptera	Ber	Larva	-do-
29.	Rice weevil	*Sitophilus oryzae* Linn.	Coleoptera	Stored grain	Grub and adult	Chewing and cutting type
30.	Khapra beetle	*Trogodemua granarium*	Coleoptera	Stored grain	Grub	-do-
31.	Grain and flour moth	*Siiotroga cerealella.*	Lepidoptera	-do-	Larva	-do-
32.	Pulse beetle	*Callosobruchus chinensis*	Coleoptera	Pulses	Grub and adult	-do-

Contd....

Common Name	*Scientific Name*	*Order*	*Host Plants*	*Damaging Stage*	*Mouth Parts*
33. Lesser grain borer	*Rhizopertha dominica*	Coleoptera	Stored grain	-do-	-do-
34. Red hairy caterpiller	*Amsacta spp.*	Lepidoptera	Polyphagous	Larva	Biting and cutting type
35. Gurdaspur borer	*Acigona steniellus*	Lepidoptera	Maize, Jawar, Sugarcane	Larva	-do-
36. Army worm	*Mythimna separata*	Lepidoptera	Paddy, Jawar, Bajra, Wheat, Maize, Sugar.	Larva	Chewing and cutting type
37. Shoot fly	*Atherigona varia soceata*	Diptera	Jawar, Bajra .	Maggot	-do-
38. Fruit sucking moth	*Ophideres* spp.	Lepidoptera	Mango, Guava, Citrus and Pomegranate	Moth	Siphoning type
39. Potato tuber moth	*Gnorinoschema operculella*	Lepidoptera Tomato, Tobacco and	Potato,	Larva	Chewing and cutting type
	0	*Agricultural Entrance*		*117*	Brinjal

Important Glossary

- Acaricide or Miticide: The chemical substances used to kill mites are known as acaricide.
- Adulticide: Chemical substance used to kill adult insects.
- Antibiosis: Antibiosis refers to the adverse effect of the host plant on the biology of the insects and their progeny infesting it.
- Allelochemicals: Allelochemicals are non-nutritional chemicals produced by an organism of one species and affect the growth, health, behaviour or population biology of individuals of another species.
- American Bolworm: Recently the American bollworm has become a serious pest of cotton in northern parts of India. It was first reported on cotton in 1977 in Punjab.
- Armyworm: The armyworm is a pest of graminaceous crops-all over the world like jawar, maize, bajra etc.
- Antifeedants: The antifeedants are chemicals which inhibit or deter the feeding of insects due to their presence on the natural food of the species concerned. An antifeedant acts by suppressing the gustatory receptors.
- Avermectins: Avermectins are macrocyclic lactones which were originally isolated in 1976 by scientists at Merck and Co, Inc from a culture of *Streptomyces avermitilis* from Japan.
- Allomone: A compound released by one organism which evokes a reaction in an individual of a various species that is favourable to the emitter but not to the receiver.
- Antimone: A substance produced or acquired by an organism that when it contacts an individual of another species in the natural context, evokes in the receiver a behavioural or physiological reaction that is maladaptive to both the emitter and the receiver.

6

Friendly Insects

Utility of Insects

Apis dorsata (Rock bee): It is found in all parts of India and big in size to another bees measuring 20 mm in length. It is a very good honey gatherer. It builds a single huge comb at the face of a rock, on branches of big trees and some times on walls and buildings. This bee is irritable and ferocious in nature. On disturbance it attacks to human beings and domestic animals and some times even death may occur. They are migratory in nature. In winter they migrate to hills and come back to plains during summer. The comb is always built in open place that's why the bees could not be domesticated up to now. The comb measures some times 1 meter across and each colony of bee is capable of storing 20 to 40 kg honey.

Apis indica (Indian bee): This bee is also found every where in India and its nature is quite different from rock bee being mild in temprament and can be domesticated easily. It is about 15 mm in length. They like to live in dark places, therefore, establish their comb in closed covered places such as hollowed tree trunks, burrows in the ground, hollowed places in walls, unused boxes etc. They build several combs side by side and parallel to each other but are smaller than the rock bee. The honey and pollen are stored in the side combs while brood rearing is restricted to the central one. On an average 3-4 kg honey is found from a comb annually. Generally they do not migrate to leave their old combs. Two types of Indian bees, *viz.*, plains type and hills type are found.

***Apis florea* (Little bee):** This is smaller than the Indian bee and found every where in India. It does not like darkness, therefore, forms its comb in the open place, *e.g.*, in the bushes, corners of roof. It builds a single comb which is very small and only 250 gm honey is obtained at a time. The honey produced by them is the sweetest one. This species could not tolerate cold therefore, is not found in hills.

***Melipona* spp. (Dammar bee):** This is the smallest species and differs from the other bees in its external appearance as well as in habits. They do not have sting. Generally, they built their comb in hollow walls or tree trunks. They construct their combs with a mixture of earth, gums and resinous substances collected from plants as they do not secrete wax to build the combs. Very little amount of honey is obtained from their combs. Economically, it is not desirable to domesticate them.

In some parts of the hills, the European honey bee *(Apis mellifera)* are domesticated. They built their comb like Indian honey bees and produce 9-10 times more honey than these. They are similar in size and structure but most industrious and peaceful than the Indian honey bee.

Business Significance

The honey bees are very useful to mankind and following are the direct and indirect advantages from them:

Honey: The honey is a colourless to light brown colour viscous fluid produced by honey bees. It contains 78% sugar (glucose and fructose), 17% water and 7% enzymes and mineral matters. To produce honey, honeybees suck the nectar from the flowers with the help of their proboscis and glossa and collect in comb. During sucking process some saliva is also get mixed to the nectar. This collected material is filled in the honey chambers and dried with the help of fanning the wings. When it is ready the mouth of honey chamber is closed with the wax. Honey is a very nutritious food having medicinal value. It has been estimated that the annual production of honey in India is about 34,01,925 kg from which the foreign money is obtained in millions of rupees.

Wax: It is secreted by the worker honey bees. Previously it was thought that the honey wax is produced by the nectar of the flowers, but now it is known that the same is produced by the wax glands situated in the ventral side of the abdomen of worker honey bees. It is secreted in the form of fluid which gets dried and remains attached in the sternites of abdominal segments. With the help of hairs present in the tarsi it is carried to the mouth where it becomes flexible due to the action of mandibles. This is used for making the comb and ceiling the honey chambers.

The honey wax is also useful to man. From this cream, polish, carbon paper and various models are prepared. It is also used in electric insulator and lithography. It has been estimated that about 1,17,933 kg of bee wax is produced annually from which lot of foreign money is being earned.

Pollination: It is known to every body that honey bees play very important role in the pollination and the gardens which are not visited by bees bear comparatively less fruits. 80 to 90% pollination is done by the bees in the plants.

Social Life, Identification and Life History: The honey bees are social insects and live in colonies with a highly organised system of division of labour. Many combs are found in a colony in which the members of the same family used to live. A family consists of 30,000 to 50,000 members which includes a queen, 200-300 drones and workers.

Queen: Every colony has a queen which is the mother and only sexually developed female. All the members of colony are the products of queen, so that it is called as mother queen. She mates with the drone, the male bee, in the air once or more in her life-time and lays eggs throughout her life. Her reproductive organs are well developed. A well developed queen is generally two or three times bigger than worker measuring 15-20 mm in length. She can be distinguished by her extended abdomen and provided with a combined sting and ovipositor. She lays 2 to 2,000 eggs per day and about 15 lakh eggs are laid in her life time. A queen lives for 2 to 5 years and when it is weak or unable to lay eggs it is replaced by one of the daughter queens.

Drone: The drone is the male bee, the main function of which is to mate with queen. It is smaller than queen and measures 15-17 mm in length. They are not permanent member's of a colony. They are driven out of the hive before the setting of monsoon and winter, therefore, die due to starvation and are raised only when new queens are reared. They are reared from the unfertilized eggs and do not possess structural modification, therefore, incapable of collecting food or defending themselves against enemies.

Worker: They are under developed females which are produced from the fertilized eggs but remain sterile due to non availability of royal jelly. Following are the main characteristics of a worker bee:

1. It is smaller in size than queen and has a sting on the terminal end of abdomen.
2. On ventral side of abdomen, wax glands are found.
3. Hind legs are modified for pollen collection.
4. The mandibles are flattened and spoon shaped which are used for molding the wax for comb building.

They are responsible for all the work necessary for the maintenance and welfare of colony such as:

1. Building of comb with wax and its maintenance.
2. To collect honey, pollen and water for the use of the colony.

3. To guard the colony against enemies.
4. To take care of the queen and feeding of royal-jelly.
5. To maintain the temperature of hive by cooling and heating the comb as per requirement.
6. To secrete the royal jelly, prepare a 'bee bread' and to distribute these foods, to grubs.

Life History

As stated earlier the eggs are laid by queen and when a colony wants to produce a new queen, the special cells are constructed at the lower border of the brood comb. On these cells, single egg is laid by the queen in each cell which hatches after 3 days. The newly hatched grubs are provided with royal jelly. The grub is fully developed in 5 or 6 days and then queen cell is capped where grub changes in to pupa and after a week adults come out by biting the cap of queen cell. The adult which comes out earlier becomes the queen daughter and it kills the remaining before their emergence.

Nuptial Flight

After 2-3 days the queen daughter takes nuptial flight accompanied by hundreds of drones during day: She overtakes her in flight. The drone which follows her takes the chance of copulation. The male dies soon after copulation and the mated queen returns to the comb. It mates only once in her life time. The seminal fluid (male sperm) is collected in a special receptacle (spermatheca) and used when required.

Oviposition

After some times the queen daughter starts egg laying and called as queen mother. She lays fertilized or unfertilized eggs at her will. One egg is laid in a cell. The eggs are long, oval and light brown in colour. They hatch in 3-4 days.

Grub

From the fertilized eggs, queen and workers and from the unfertilized eggs drones are born. The grubs are cylindrical and light yellow in colour, They are fed with royal jelly for 2-3 days after that they are provided honey and nectar etc. The grub period lasts for 5-6 days.

Pupa

Full grown grub forms a cocoon and pupates inside the cell. The pupal period lasts 7-14 days depending upon the type of adult to be produced. The time required for different types of adult in different stages is given below:

Adult	*Egg*	*Grub*	*Pupa*	*Total*
Queen	3 days	5-5.5	7-7.5 days	15-16 days
Worker	3 days	5-6 days	10-10.5 days	18-28 days
Drone	3 days	6-6.5 days	14-14.5 days	23-24 days

Activities During the Year

The honey bees remain active generally through out the year except during severe winter. Following are the main activities of honey bees:

Combing: The comb of honey bees comprises of several hexagonal cells on both side of mid-rib. They built their comb to the ceiling of the cavity or lower side of the branch. In the upper cells they store honey and pollen and in middle and lower raise their young ones. Freshly built comb is generally white, but becomes dark after some times. The filled cells are closed with a cap.

Swarming: During spring and summer when conditions are favourable and food is available in plenty, the bees multiply greatly with the result the comb becomes crowded and the bees begin to make preparations for swarming. At this stage the queen daughter cells are built at the bottom and when new queen is ready to emerge out, the new queen and a large number of workers which have previously filled the cells with honey, leave the nest to start a new colony. Swarm settles in a suitable place already searched out by the workers for building new comb.

In a parent colony the first queen daughter which emerges after swarming, kills the baby queen in the other cell and establish herself as a queen mother. After that, they start their routine work of gathering nectar and pollen.

Migration or Absconding: The following are the main reasons for the migration of bees:

(i) Scarcity of food.

(ii) Unfavourable environment.

(iii) Destruction of combs by man or other animal.

When they migrate from one place to another they take all honey nectar with them from the old comb. But on disturbance, they fly to save their life and will return to same place after some time. Then they start to search a suitable place.

Air-Conditioning

Among the living creatures honey bees are the only organisms which make their comb air-conditioned. They keep their comb warm in winter and cool during summer. During winter they generate heat by sitting on one another and rubbing their legs due to which the temperature of the comb remains about 20° F, more than the environment. In summer they start fanning with their wings with the result water evaporates from the honey and comb remains cool.

Apiculture or Bee-keeping

Rearing the bees in artificial hives is known as bee-keeping or apiculture. In India the bee-keeping industry was started about 50-60 years ago, however, the references are available in old literature (Mahabharat). In the old age, people after wrapping the blanket on the body or after smoking in night, collected honey from the comb. This was a crude method. After some times people thought to keep the honey bees and many villagers took interest in keeping honey bees and provided various types of hives in their houses. The hives were not much different from their natural comb.

These were hollowed logs, empty boxes, earthen pots and walls etc. In these indigenous methods, it is mostly left to chance whether a hive gets inhabited or not. It was not possible to have any control over the activities of bees. They can not be checked from robbing and swarming etc. When honey was extracted, a large number of bees were killed and comb built by honey bees was destroyed. The honey obtained was not pure and contained brood juices, cocoons, pollen grains and parts of the body of the bees, with the result modern method of bee-keeping was developed.

Modern Method

The artificial comb was first introduced by Langstorth in 1851 in America., In India attempts were made during 1882 in Bengal and 1884 in Punjab for artificial bee-keeping but success could not be achieved due to the lack of scientific know how. From 1907 to 1919 researches were made on bee-keeping under the guidance of Imperial Entomologist and work was started during 1931 in Madras, 1933 in Punjab and 1938 in Uttar Pradesh. At present in every state of the country there is a bee-keeping centre and about 15 lakh rupees are earned every year.

The following apparatus and equipments are required for bee-keeping:*(i)* Bee hive *(ii)* Comb foundation *(iii)* Smoking apparatus *(iv)* Hand gloves *(v)* Scraper *(vi)* Net *(vii)* Knife *(viii)* Honey extractor apparatus.

Bee Hive: Various types of bee hives are available for bee-keeping. They are wooden boxes having two parts upper 1/4 comb chamber and lower 3/4 brood chamber. In between these two chambers, a net (series of parallel wires) is used by which worker may go from one chamber to another but not the queen. This net is known as queen excluder. The queen is kept separate in the lower chamber only so that she may not lay eggs on the upper chamber.

The bee hive remains closed from all sides and is provided with a small hole on the bottom from which only a worker may pass. Since the Indian honey bee by nature makes parallel comb in a place, therefore, wooden planks are hanged at a distance of 10 to 12.5 cm in the lower chamber. Each two planks are attached with a wire on which comb foundation is placed. The following two types of bee hives are more popular in India:

(i) Ghos box—36 cm × 21.5 cm

(ii) Newton box-20.2 cm × 14.0 cm

Some other familiar boxes are as follows:

(i) Langstorth box (American hive) 42.2 cm × 31.1 cm

(ii) Pant, Kanje and Jeolikote No.1 42.2 cm × 12.3 cm

(iii) Dadant box (Russian hive) 47 cm × 28.6 cm

(iv) Thompson box 30.5 cm × 15.2 cm

Comb Foundation: It is a sheet of pure bee wax with the base of the worker brood cells embossed on it. It is fixed in a frame with wires and the bees construct a complete comb on it. It helps in obtaining regular strong worker brood cell comb.

Smoking Apparatus: It is a tin container from which smoke is generated when bees are furious, which become silent due to smoke.

Hand Gloves: These are rubber gloves and used at the time of handling the bees.

Scraper: It is a flat iron plate used for scraping the floor of the box.

Net: It is made up of either silken or cotton thread and wore on head, to get protection from their attack.

Honey Extracting Apparatus: Honey extractor works on the principle of centrifugal force and throws out honey from the cells without breaking the combs.

Knife: knife is used for decapping the cells in which honey is filled so that it may be extracted easily.

Catching of Bees

Now-a-days hives and bees including other accessories are made available by various agencies for purchase. But some people still capture wild colony and domesticate them. The wild colonies of Indian bees are usually found in burrows of trees, cavities of walls etc. and are easily procurable in the swarming and migrating seasons. The availability of food supply should be provided.

Natural Enemies of Bees

Honey bees suffer from a number of diseases and have several enemies.

Diseases

The bees generally suffer from nosema, acarine septicaemia and amoebic, fungus and paralysis diseases. These are caused by a protozoan *Nosema apis,* a *mite-Acarapis woodi,* a bacterium-*Bacillus apiscepticus,* an amoeba-*Vahlampfia, mellifica* several fungi-(Mucor spp, *Aspergillus* spp and *Mycoderma* spp etc.) and virus, respectively. Due to these diseases they are unable to fly, crawl on the ground, climb on grass and collect in small groups in front of the comb.

Bee Enemies

The following are the important enemies of bees:

Wax Moth: There are two species of wax moths namely greater wax moth (*Galleria mellonella* Linn) and lesser wax moth *(Achroia grisella).* The caterpillars of these moths make silken tunnels in the comb under the cover of which they feed upon bee wax. During severe infestation the comb is covered with silken tunnels and faecal matter of the caterpillar due to which bees become incapable of guarding against them and they abandon the nest.

Sphinged Moth (Acherontia styx): It is called as honey robber. It sucks the honey from the comb with the help of proboscis.

Black Ant (Camponotus compressus): They also eat honey from the comb and are particularly trouble some after rain.

Wasp (Vespa spp): Five different spp. acts as predators of bees at the entrance of the hive or in the fields.

Birds: The green bee-eater (*Merops sp.)* and the kings crow (*Dicrusus spp.)* catches the bees during flight and eat them.

Protection Measures

Following measures may be adopted to protect the bees:

1. The entrance door should be small so that *Acherontia styx* may not enter.
2. The combs should not be left open to protect them from wax moths. If there is an attack, wax should be removed and bees should be provided with new comb.
3. The bees suffering from paralysis should be destroyed and spraying of sulphur may be done on the comb.
4. The following mixture may be used in case of bees suffering from Acarine —

Safral oil	—	1 part
Nitrobenzene	—	2 parts
Petrol	—	3 parts

Silk-moth

Scientific name	:	*Bombyx mori,* Linn.
Order	:	Lepidoptera
Family	:	Bombycidae

Distribution

China was the pioneer to have a know-how about sericulture industry. At present, it is a well established industry in many countries of Asia, Europe and America with particular reference to Japan, Italy, France, Spain, India and. Mayanmar.

Economic Importance

Silk is the another valuable product which we get from the insect. We get food (honey) from honey bee and cloths from silk worm. In the ancient age China has sold the silk in the value of gold. Todays, also, it is an important product besides cloth, the parachutes are also prepared from this which are used during war etc. The oil is extracted from the cocoons of tasar silk worm and used in the preparation of some important medicines.

Silk

Silk is the secretion from the salivary glands which are found on both sides of the alimentary canal of silk worm larvae and this secretion hardens into fine threads called silk. The cocoons with which pupae are covered by the worms are utilized for silk production. Most of the Lepidopterous caterpillars and some other insects spin silk, but those of only two families viz., Saturnidae and Bombycidae, produce silk of practical use on account of quantity and quality. Silk is spun in one single reelable thread which is collected and used for manufacturing the threads and cloths etc. According to some people India is be considered the origin of this insect while others considered China as a native place. Princess Lotzon of China investigated the process of making silk as early as 269 BC. China made a tremendous progress and started its export to various countries during 2000 BC. The source and method of obtaining the silk was kept secret about 2500 years. The government made a strict rule that the person who will leak the secret to any of the foreign countries will be hanged. But still two religious bishop who went there, learnt the know-how about silk worm and collected the eggs of silk worm in hollow bamboo pieces and carried to Kuptuntunia, from where it spread to the European and Asian countries.

Species of Silk Worms: Five silk worm species are found in India which are described as below:

Tassar Silk Worm (Antherae paphia, L): It is generally found in the forests of Bengal. Assam and Uttar Pradesh and is not domesticated. The important host trees of this silk worm are sal (*shorea rsobusta)*, ber *(Ziziphus mauritiana)*, Asan *(Terminalia iomentosa)*, and Arjun (*Terminalia arjuna)*. The cocoons are hard light brown to dark in colour and have to be collected from the jungle. They are reeled and if pierced are spun. The another species of the silk worm *(Antheraea toyley)* is found in the tarai region of Himalaya on Oak, the cocoons of which are hard and white in colour.

Eri Silk Worm (Philosamia ricini, B): This silk worm was mainly found in the forests of Assam but now also in Bengal, Bihar, U.P., Orissa and Chennai. As its scientific name indicates, it feeds on castor *(Ricinus communis)* leaves and domesticated all over the country where castor is grown. Its cocoons can not be reeled, therefore, have to spun and inferior in quality, cocoons are white in colour.

Muga Silk Worm (Antherae assamia, H): This silk worm is found in Bihar, West Bengal, Orissa and Assam and can not be domesticated. The caterpillars feed on the leaves of cinnamon and matchilis plants. The cocoons are reeled and are of brilliant yellow colour from which silk of golden yellow colour is obtained.

Deo Muga Silk Worm (Thiopaila religiosae, H): It is found in Uttar Pradesh and Tarai region of Himalaya. The caterpillars mainly feed on the leaves or machilis and Ficus plant. Its thread is generally used for making the fisheries net.

Mulberry Silk Worm (Bombyx mori, L): It is the most important silk worm and reared extensively in various part of India. Its caterpillar feeds on mulberry leaves. The cocoons are white in colour from which best quality of silk is obtained. There are several races of the mulberry silk worm which may be grouped in to two categories, *viz.*, univoltine and multi-voltine. The univoltine has an annual life-cycle while multi-voltine race passes through many generations in a year. The multivoltine race produces inferior quality of silk and is reared in Bengal, Assam and Mysore.

Life History of Mulberry Silk Worm

The silk worm is moth of creamy white colour measures about 30 mm in length and with a wing span of 40-50 mm. The male is smaller than female. The head is small and bears a pair of black compound eyes and bipectinate antennae. The mouth parts are vestigeal, therefore, does not take food and lives only 2 to 3 days. The anterior portion of thorax is narrower than the posterior. The fore wings are provided with dirty dark coloured stripes and the body is covered with hairs. The moths copulate by keeping their mouth to opposite direction. They mate immediately after emergence during May-June.

Egg

Eggs are laid in the night in clusters on the under surface of the mulberry leaves. A female lays about 300-400 eggs popularly called as silk-seeds. The eggs are small, pale white and seed like in appearance. At the time of hatching they become black and hatch in 10-12 days during summer and 30 days during winter. In the univoltine race; the eggs do not hatch during winter but remain in hibernation, therefore, one generation is found in a year in such cases. Such type of race is found in European countries and Kashmir and Punjab in India. Contrary to this 2-7 generations are found in multivoltine race.

Larva

The caterpillar on hatching, is white to dark in colour and about 3 mm in length. There are 3 pairs of thoracic legs and 5 pairs of abdominal legs which are provided on the 3, 4, 5, 6 and 10th abdominal segments. A horny appendage known as dorsal spine is found on the dorsal surface of eight abdominal segment. Young caterpillars are reared in trays on tender mulberry leaves at 25-27°C. Each day, the feed is to be given 3-4 times with small quantity of leaves. After 5th day when they become big, a net of small mesh is placed over them and the succulent leaves are provided over the net. The larvae on getting the food crawl above through the meshes of the net.

After about 2 hours, the net is lifted and placed on another tray. The caterpillars along with the unused mulberry leaves is removed and thrown away. The larva moults 4-5 times after every 6-7 days and becomes mature in 30-35 days. In cold countries it takes many months to become full grown. The full grown caterpillar is creamy white in colour and about 75 mm long. The matured worms are picked up and kept in cocooning baskets. They make cocoons generally within 25 hours.

Pupa

The cocoon is 38 mm in length and 19 mm in breadth, oval in shape and white or yellow in colour. The larva pupates in side the cocoon which is made up of a single thread. It spins about 15 cm thread in a hour. The pupa inside the cocoon is reddish-brown in colour and measures 25 mm × 7 mm long. The pupal period lasts for 10-15 days. At the time of emergence of adult, it secretes an alkaline fluid which pierces the cocoon and it comes out. To obtain good quality of silk the moths are not allowed to emerge but the pupae are killed inside the cocoons either by sunheating or putting them in boiling water after 10 days of cocoons formation. This process is known as stiffing.

Immersing in hot water softens the cocoons by dissolving its gummy material and therefore, one end of the silk thread can be easily teased out. Those cocoons which are intended for further rearing are kept separately in a cool place and moths are allowed to emerge and lay eggs.

It has been estimated that about 40-50 thousand caterpillars hatch out of 28 gm. of silk seed and they require about 337 to 406 kg of leaves during their development.

Sericulture

The sericulture industry in India is about 1000 years old and 10 million kg of silk is produced annually, the 50% of which is produced in Mysore state. Besides, it is also produced in West Bengal, Chennai, Kashmir, Assam, Bihar, Maharashtra,

Madhya Pradesh. U.P. and Orissa. The Mysore race of silk worms and Nistary Bengal are the best. During second world war period many schemes have been started for the development of Sericulture Industry. In 1948 the Central Sericulture Research Station was established at Baharampur (West Bengal) for the purpose of improving the quality of silk worm and ultimately the quality of raw silk. For the further development of the silk industry in India, in 1946, the Central Government has set up a statutory Central Silk Board, U.P. Government has set up a sericulture Research Station at Dehradun in 1948. The major mulberry raw silk producing states are Mysore, West Bengal, Kashmir and Chennai. Assam produces a large quantity of raw silk of non mulberry type. The following points may be followed to produce good quality of silk.

Stiffing

As the silk is spun in one single reelable thread, it is liable to cut into pieces, if the moths are allowed to emerge. Therefore, after about 10 days of cocoon formation, they are kept in boiling water or exposed to hot rays of sun to kill the pupae inside and thus prevent the cutting of the silk thread in to numerous pieces.

Selection of Cocoons

After killing the pupae inside, the cocoons are selected and separated on the basis of their colour and brightness.

Reeling and Cleaning of Thread

Immersing in the hot water softens the cocoons and the silk thread may be teased out easily. After removing the external fluffy layer continuous filament from, the middle layer is reeled with the help of suitable machines. Such threads from 4-5 cocoons are put together through a hole known as eyelet or guide of a reeling machine to form a single thread of reasonable thickness. This is a raw silk and to make spun silk, these threads are again boiled in water and cleaned by washing them in acidic solutions. Many such threads are reeled out together and finally fibre silk is produced.

Materials for Sericulture

1. *Benches*—They may be of wood or bamboo.
2. *Trays* —These are made of split bamboo and may be of various shapes and sizes.
3. *Nets*—These, are small nets of cloths by which the leaves and excreta of larvae is cleaned.
4. Leaf cutting knife
5. Hygrometer
6. Heater and cooler to maintain the temperature.

Before establishing a sericulture industry in any place it is essential to plant the mulberry trees so that the leaves may be obtained continuously. The caterpillars are reared in rooms, therefore, they may be situated in open place and provided verandah around them. The rooms should be clean and have a good ventilation. The legs of the benches should be kept in water so that ants may not approach the caterpillars. The trays having caterpillars are put on these benches for rearing.

Diseases of Silk Worm

The silk worm suffers from a number of diseases. The important ones are given below —

Pebrine: It is of two type:

(a) *Viral Pebrine:* This is caused by the virus namely *Borreline botnbycis,* the larvae suffering from this disease are killed 8 to 10 days of infection. The diseased caterpillars may be separated and all the apparatus be washed with 30% solution of trichloro acetic acid for 15 minutes and then with water.

(b) *Protozoan Pebrine:* It is caused by a protozoan namely *Nosema bombycis.* Both larvae and adults suffer from this disease. The attacked moths shrink and become irregular while the caterpillars remain shorter, pale and translucent in colour, shrivelling and died before pupation. The disease is transmitted through egg from pebrinised mothers. As a protective measure, eggs from healthy females may be used for further rearing. Disinfection of rearing room and appliances with formalin and destruction of infected moths and worms should be followed to check the spread of disease.

Grassrie: It is the bacterial disease and the attacked caterpillars swell and their body fluid becomes cloudy and thick. The skin at times bursts. Care should be taken that the larvae should not be fed on mature leaves first and tender leaves afterwards.

Flacherie: The diseased caterpillars become thin and give out green fluid from their mouth. The caterpillars die and their bodies give an offensive smell. This disease is generally caused by indigestion, therefore, over feeding should be avoided. The other precautions like supply of wet and fermented leaves or dusty and muddy leaves, over crowding of larvae in trays should be avoided.

Muscardine: This disease is commonly found during rainy-season and caused by fungus. The suffered worms become soft and die and dead worms become stiff after some times. Since the disease is infectious, the diseased larvae should be burnt and general cleanliness of the rooms and appliances should be practiced.

Lac-Insect

Scientific name	—	*Kerria (Laccifer) lacca*, (Kerr)
Order	:	Hemiptera
Sub-order	:	Homoptera
Family	:	Lacciferidae

Distribution

Besides India, it is found in Pakistan, Mayanmar, Sri Lanka, Jawa, Malaya, China, Indo-china and Thailand.

Host Plants

The lac insect is domesticated on the following plants:

1. Palas *(Butea monosperme, B. frandosa)*
2. Ber *(Ziziphus mauritiana, B. Jujuba)*
3. Kusum *(Schleichera oleosa, S. trijuga)*
4. Ghont *(Ziziphus xylopyra)*
5. Jallari *(Shorea talura)*
6. Arher *(Cajanus cajan)*
7. Pipal *(Ficus religiosa)*
8. Grewia *(Grewia spp.)*
9. Babul *(Acacia arabica)*
10. Khair *(Acacia catechu)*

Economic Importance

India is the major lac producing country accounting for nearly 80-85% of the total world production and foreign exchange of 10-15 crore rupees is earned annually. Over 90% of the country's production (which is about 5 lakh quintal for crude stick lac) is exported to foreign countries in the form of shellac seedlac and other refuse lac etc. Thus it is an important cash crop and is a source of living of many tribes.

In the ancient age the people of India knew about its uses. The literature on lac is available in Arthary Ved and Mahabharat which indicates that lac culture was in advance stage during that period. During ;1590, the Aine-Akbari was published in which there was a mention about lac polish. Similarly in 1676 Tabneer has written that the colour of lac is used in printing in India.

At present it is used in a very large number of industries, the chief among them being the gramophone record, furniture and floor polishes, paints and varnishes, electrical goods, sealing wax, printing ink industries, etc. Besides, also used in making bangles, toys, wood work, silvering of back mirror, as gelatine in stiffening hats and filling of ornaments etc.

Looking to its importance, India Government has started research work on lac in 1921 and established Indian Lac Research Institute during 1925 at Nankum, Ranchi (Bihar). Indian Laccess committee was founded during 1930 to have a check on its export and to frame rules etc.

Lac

Lac is the valuable resin secreted by the lack insect. The lac insects feeds on the sap of certain plants and continuously secretes the resin as protective covering throughout its life. The lac glands open on the dorsal surface of the body through various holes. Besides, three wax glands are also situated on the mouth, at the meso thoracic spiracle and around the anus.

The wax is secreted by these glands which protects to harden the lac on these parts. In absence of this, female may die. Lac is the only one of its kind in its origin as an animal resin and possesses certain peculiar properties that make it a very versatile multi purpose resin. The following constituents are found in the lac:

1.	Resin	68%
2.	Coloured materials	10%
3.	Wax	6%
4.	Gum	5.5%
5.	Sugar	4%
6.	Debris	6.5%

Identification and Life History of Lac Insect

The adult male and female are different from each other. Female is about 3 times larger than the male.

Male

They are pinkish-red in colour and are of two types-winged and wingless. Winged male possesses only one pair of translucent membranous fore wings. They are mostly found during dry season (Baisakhi and Jethawi). They survive only 3-4 days and die after copulation. The following are the main characteristics of these:

1. The head is a large, prominent mouth parts but vestigeal.
2. Ocelli are 2 pairs with seven segmented, antennae which have hairs on them.
3. Tarsi 3 segmented.
4. The abdomen is 8 segmented, broader anteriorly and narrower posteriorly. The last abdominal segment bears pointed penis.

Female

The female is pinkish in colour and about 1.5 mm in length. The ventral surface of the body is flat while dorsal surface is convex. The main characteristic are given below —

1. The female deprives of eyes, legs and wings.
2. Antennae are vestigeal small and 3-4 segmented.
3. The mouth parts are piercing and sucking type and rostrum is two segmented.
4. The female remains attached at a point and sucks the juice from the host.
5. Meso thorax is provided with appendage on which spiracles are open.
6. The abdomen is round and on the dorsal surface, a spine is provided. The last abdominal segment is two segmented and anus is friged.

Life History

Three developmental stages are found in its life cycle, *viz.*, egg, nymph and adult.

Egg

The eggs are laid by the female within the lac-cell. It lays about 200-500 eggs which may be fertilized or unfertilized.

Both types of eggs give birth to male and female nymphs. The female lays three types of eggs—

1. The eggs from which both males and females hatch out in equal number.
2. From which males are less than females.
3. Males are more than females.

An individual egg is 0.4 mm × 0.2 mm long and pinkish in colour which changes into brown after sometimes. The female generally lays well developed eggs which hatch within a few hours, that's why some scientist call them viviparous females.

Nymph

The newly hatched nymph is 0.6 mm × 0.25 mm long and pinkish in colour. It possesses a pair of antennae, three pairs of legs, two compound eyes and six anal setae. The nymphs emerge in large numbers and crawl over the surface of twigs and branches of plants in search of suitable pieces for settlement. They mostly settle on shoots gregariously.

Once settled, the nymph thrusts its proboscis into the bark of the twig and does not move from its place. After one or two days of settling, they start secreting resin from glands distributed under the cuticle all over the body, except near the mouth parts, the two breathing pores and the anus. Thus the nymphs gets encased in a cell of its secretion which increases in size with the growth of the insect. They moult thrice before reaching maturity and after Ist moult the nymph loses its eyes, legs and antennae but anal setae become 10 instead of 6.

The sex is readily recognized by the shape of lac cells even in the early stage of development. In the case of the male cell, the growth is more along the longitude while in the female cell it is more along the vertical axis. The differentiation between sexes is more pronounced after the first moult. The male cell is elongated with a pair of holes at the anterior end from which thin whitish hairs comes out, while a female cell is some what globular with an irregular margin and with three holes on the top, one at the hind end and two at the anterior end.

The male nymph develops the organs after second moult, both the males and females continue to develop. From the male cells both winged and wingless males emerge after about two months. These males fertilize the females in their cells and after fertilization female develops very fast, swell up and secrets the lac at a faster rate.

The size of the female as well as that of the enveloping lac cell increases rapidly and reaches a size several times that of the male lac cell. Thus the females are the chief source of lac secretion. As the time for egg laying approaches, the posterior region of the female develops a yellow spot which is indication of egg laying. The lac insect usually passes through two generations in a year.

Lac Culture

The lac industry is well developed in India and at present it accounts tō about 70-80% of the total world production. The lac produced by different states of the country is given below:

Bihar	41.3%
Madhya Pradesh	26.5%
West Bengal	20.6%

There are two distinct strains of the lac insect in India, called Kusumi and Rangeeni. The former is raised mainly on the Kusumi trees or on other hosts with Kusumi brood and the latter on all other lac hosts such as palas, ber, ghont and arhar etc. Each of the strains produces two crops in a year, thus there are four crops which have been shown in the following chart:

Crops

Kusumi	Rangeeni		
Aghani	Jethawi	Katki	Baisakhi
(6 months)	(6 months)	(4 months)	(8 months)

Starting	Harvesting	Starting	Harvesting
June-July	Jan.-Feb.	June-July	Oct.-Nov.
Starting	Harvesting	Starting	Harvesting
Jan.-Feb.	June-July	Oct.-Nov.	June-July

Of the countries total production; 90% lac is obtained from the Rangeeni cròps and only 10% from the Kusumi crops. Though Kusumi crop is produced in smaller quantities but superior in quality. The production of all those crops is given below:

Baisakhi crop	2,91,600 quantals
Katki crop	1,12,200 quantals
Aghani crop	34,400 quantals
Jethawi	15,400 quantals

The two methods of lac cultivation are popular:

1. Local cultivation
2. Modern method

Local Cultivation

This is the crude and unscientific method of lac cultivation in which lac is collected from the trees growing in jungles. Lac is left for continuous self-inoculation on the same tree and every time a partial crop is taken. The method of lac cultivation has generally failed to provide sustained supplies of brood lac for crop inoculations and optimum yield of lac.

Modern Method

This method was developed by Indian Lac Research Institute, Ranchi and is known as coupe system. In this system, as against the local practice of keeping all the tree under continuous cropping a certain number of trees in a coupe or compartment are inoculated fully and when lac matures it is fully reaped and another set of trees from another coupe is taken up for subsequent inoculation. The coupe from which the crop was reaped not only got sufficient time to put forth new shoots and foliage but also get rest in regaining the vigour. To get good crop, following points should be kept in mind.

Selection of Suitable Site: The site should be selected on the basis of host plants, where environmental conditions, suitable for host trees should be preferred. The excessive heat and cold should be avoided and rainfall may be about 60.

Pruning of Host Plants: This is an important and essential operation carried out with the object of inducing the tree to put forth maximum number of shoots suitable in condition, for successful colonisation by the larvae of the lac insect. Before starting lac cultivation, care should be taken that the tress have sufficient tender branches to enable the lac insect to establish itself. If it is not so, the trees should be pruned. Pruning should always be carried out lightly and branches, more than 2.5 cm in diameter should not be cut. The pruning is dependent upon the type of the host plant, *e.g.*, in case of ber trees, a large number of succulent shoots are produced after pruning while palas and Kusumi do not ordinarily require pruning.

Inoculation of Brood Lac: Lac sticks bearing alive mother cells which produce lac larvae are known as brood lac. The brood lac sticks should be cut in to convenient length of 15-30 cm and tied in succulent shoots of the host tree. The tying may be done either longitudinally or laterally. The correct amount of brood lac (covered 15 to 20 times of its length) should be used for inoculation. The larvae after hatching will transfer themselves into succulent shoots. The brood used for inoculation should not be left on the trees for more than 3 weeks because keeping beyond this time, there is a danger of enemies to spread. When the brood lac is removed from the trees, it is known as Phunki lac. There are following types of lac:

Ari Lac: This is the immature lac and cutting of ari lac should be avoided.

Stick Lac: Mature lack harvested in the form of sticks is called as stick lac.

Seed Lac: The lac obtained after removing and washing from the sticks is known as seed lac.

Dust Lac: Dust lac is obtained after grinding the seed lac.

Shel Lac: Shel lac is prepared after heating the seed lac and dust lac.

Scrapping of Brood Lac: When the brood lac sticks are removed from the trees (Phunki lac), all other, either dead or rejected as unfit for brood purposes, is scraped as early as possible. The possible date of crop maturity may be known by the following points:

1. Appearance of cracks in the lac before 15-20 days.
2. Granular appearance of the body contents of the female lac insect.
3. Eggs become separated from the masses.
4. The eggs seem to be granular.
5. The lac becomes dry and yellow in colour.

Washing of Lac: The lac obtained after scrapping is washed with water and dried under shadow. The colour (lac dye) is obtained due to the washing which is used for various purposes.

Preparation of Shel Lac: The following are the methods for preparing the shel lac:

(i) Indian method

(ii) Hydraulic method

(iii) Solvent method

(iv) Autoclave method

Indian Method

It is the oldest method in which the seed lac is put in to long cloth bags of 6.5 cm width. The bag is tide with a pivot of wheel and circulated. The heat is also provided side by side. The melted laç due to the pressure comes out which is collected and made in the form of tablets. This is an expansive method, it is estimated that to wash 1 quantal of lac required 1 meter cloth, 1 quantal coal and 6 men in a day.

Hydraulic Method

This is a machine in which the seed lac is put in a cloth on a which a spray of water is provided. The bottom of the machine in heated with stem by a pipe to get the lac melted. The lac is pressed by a plunger operated by hydraulic pressure and collected. To remove the moisture, the lac is heated in another machine and then stored after cooling. This is also an expansive method and can not be used in small scale.

Solvent Method

The seed lac is dissolved in spirit and separated by distillation process. The spirit may be used for further washing. The main draw back in this process is that after dissolving the second time, the spirit can not be separated by distillation hence, there is chance of fire.

Autoclave Method

This process has been developed by Indian lac Research Institute, Ranchi in which the seed lac is put in the bags of iron nets and heated in a special autoclave. The lac is collected on the lower surface of the autoclave. After some time it is removed from here and heated again. The melted shel lac is spread in to thin sheets and dried.

Enemies of Lac Insect

The enemies of lac insect may be broadly grouped in to two:

(i) Insect parasites and predators,

(ii) non insect enemies.

Parasites

The parasites of lac belong to the super family Chalcidoidea which lay their eggs inside the lac cell either in or on the body of the lac insect. The grubs after hatching, feed on the lac insect and the adults come out by biting a hole in the lac cell. Five to ten per cent damage is caused by these parasites.

The important species of parasites are as follows:

Coccophagous tschirchii Mahd.

Tetrastichus purpureus. (Cam.)

Mariettas javensis How.

Eupelmus tachardiae How.

Predators

The two predators namely *Eublema amabilis* and *Holcocera pulverea* are more important. The larvae of these moths feed on the lac insect as well as cell of the lac. The three species of *Chrysopa* also attack the lac insect, the larvae of which suck the body fluid of the host. The damage caused by the predatory insects is estimated to the tune of 40%.

Non-insect Enemies

Some birds and squirrels also destroy the lac insect and lac encrustations and caused considerable loss.

7

PEST MANAGEMENT

Rapid progress in agricultural production technologies like introduction of high yielding hybrids and varieties, increased use of chemical fertilizers, better management of irrigation, mixed and relay croppings etc. has considerably changed the agro-ecosystem.

It has resulted in cropping up numerous problems. Amongst them insect pests are proving major constraint in achieving expected yields of high yielding hybrids and varieties. Till recently use of insecticides was considered most effective tool to overcome this problem.

However, indiscriminate use of insecticides has led to serious consequences like harmful residues, pest resistance to insecticides; pest resurgence and outbreaks of secondary pests. This has brought a complete change in strategy of pest control.

Single factor approaches are inadequate hence discarded and replaced by a universally accepted concept, integrated pest management (Stern et al. 1959; Rabb and Guthrie 1970; Luckman and Metcalf 1975).

Pest Control Objectives

The main objective of integrated pest management is to optimize and not maximised pest control in terms of overall economic, social and environmental values. This approach is valid even under diverse production technologies. To get economic production it is essential to combine all suitable techniques and methods of pest suppression in as compatible a manner as possible to maintain pest populations at levels below those causing economic injury.

Such programmes will require an intensive knowledge on the key pests along with those of other minor pests, leading to economic damage threshold; an effective pest forecasting system; critical knowledge of biological factors regulating the population fluctuations of concerned pest species, and finally detailed knowledge of possible interaction between different control measures.

Perception of Pest Control

The concept of integrated pest management was first propounded by Geierr and Clark (1961) who suggested that all available techniques should be evaluated and consolidated into a unified programme to manage pest population so that economic damage is avoided and adverse side effects on the environment are minimised. The integrated pest managements is now widely recognised as the intelligent selection and use of pest control actions that will ensure favourable economic, ecological and sociological consequences.

The definition given by FAO Panel of Experts (1966-72) is that it is "a pest to management system that, in the context of the associated environment and the population dynamics of the pest species, utilizes all suitable techniques and methods in as compatible a manner as possible and maintains the pest population at levels below those causing economic injury."

Geierr: (1966) has laid down the following basic principles for the practice of pest management:

1. Determine how the system of pest needs to be modified to reduce its numbers to tolerable levels, that is below the *economic threshold*.
2. Apply biological knowledge and current technology to achieve the desirable modification, that is *applied ecology*.
3. Devise procedures for insect pest control both suited to current technology and compatible with economic and environmental quality aspects, that is, *economic and social acceptance*.

Apple and Smith (1976) have detailed the problems and prospects of integrated pest management and Brader (1978) is of the view that future development in agriculture production will depend largely on integrated pest management system. Teetes *et al.* (1980) have given *a* scheme as to how individual practice of control may be fitted together into an operational and comprehensive integrated pest control programme in sorghum. Flint and Van den Bosch (1981) gave basic insect pest management concepts (Aggarwal *et al.* 1983, Prem Kishore 1990, Prem Kishore and Gahukar (1992).

The concept of integrated pest management is dynamic and has come to stay. It may change with the advancement of our knowledge of crops, their key pests and control. The concept though appears simple in theory but there are innumerable problems to evolve such schedules. It is a phased system and involves following steps:

(i) To develop an integrated control programme against any pest it is essential to determine possible effective methods for its control.

(ii) To establish economic threshold and economic injury level at which the control operations should be initiated.

(iii) To bring about sysnthesis of possible effective control methods in such a way that the integrated components should hold the pests below economic injury levels and avoid disruption to environment.

Economic Injury Levels and Economic Threshold: *Economic injury levels (EIL)* indicate the pest densities, *i.e.*, numbers of pest per unit area, at which control measures are economically justified. At these levels the cost of control is less than the loss farmer or other resource producer would suffer if control action were not taken. The careful determination and use of these levels is essential for pest management and maintenance of environment.

Economic injury levels tell the farmers or producers the maximum number of a pest their crops can tolerate but do not indicate when to do some thing about the problem. Therefore, a second population density level known as *economic threshold (ETL)* has been devised. This can also be called as *control action threshold*. Economic threshold tells the farmers or producers when they must take control actions to prevent an impending pest out break. As the economic threshold is lower than the economic injury level, it allows enough time for control measures to take effect before the population size and resulting damage reach the economic injury level.

The concept of economic threshold was first described by Stern et al. (1959) and Stern (1973), as the density at which control measures should be initiated to prevent an increasing pest population from reaching the economic injury level. However, it is not too simple, particularly in view of recent models put forward by Hillebrandt (1960); Headly (1972) and Hall and Norgaard (1973). Headley (1972) stated that economic threshold is that population level where the marginal benefit from damage prevented by control programme is equal to marginal cost of realizing that population through a control programme. The basic elements as proposed in his model are: (i) a pest population growth function, (ii) a pest damage function, (iii) a yield function, and (iv) a pest control function. Hall and Norgaard (1973) indicated that Headley's model is static one and gives no information on optimal timing of application of control measures during the season. They showed that the economic threshold varies over time and under certain assumptions increases with time; so that the closer the harvest time, the higher is the level of pest population that will be tolerated before control are applied. They suggested following elements in their model; a pest population growth function; a pest population kill function; a pest population damage function; a product yield function; and a pesticide cost function.

As the economic injury level may vary from area to area, season to season, crop to crop, stage of the crop or with man's changing scale of economic values, the concept of economic threshold will also undergo a change. Thus it can be said that this concept is still in the process of evolution.

Monitoring and Sampling

Monitoring is pre-requisite for the success of insect pest management programme and for assessing the prevalent insect pest situation, *surveillance* of insect pests is needed. Periodic *intensive* and *extensive surveys* will establish the correlation between pest behaviour and the weather, the cropping pattern and prevalent agronomic practices, Monitoring is to establish economic thresholds; to evaluate influence of weather, natural enemies or other natural suppression factors; to gauge the effectiveness of insect pest control action, or to determine the most damaging stage of the pest. (South Wood 1966; South Wood and Norton 1973; Luckman and Metcalf 1975).

Effective monitoring will come from sound sampling procedures and careful watch on ecosystem. A constant monitoring is essential because the pest population will fluctuate with changes in environment. The average pest population over a period of time is referred to as the general equilibrium position. If natural enemies are few, the pest population will increase.

The most important sampling procedures are random sampling, point sampling, trap sampling and sequential sampling. In *random sampling* counts of pest numbers, damage, or natural enemies are taken at random spots in a field. Different sampling spots should be chosen each time the field is surveyed. Under *point sampling* pests and natural enemies are counted alongwith a more detailed monitoring of crop maturity state in one or more areas of the field. *Trap sampling* is employed for detecting the actual presence of pest species, first appearance of migrating insects, or the emergence dates of adult egg laying stages. Among the various traps, light traps, sticky traps and pheromone traps are common.

The most potential ones are *pheromone traps.* This type of traps exploits the chemical olfactory stimulants (pheromones) insects use in communication. *Sex attractant pheromone traps* have proven extremely useful in giving information about activity of peaks in various pest species. It helps in determining the advantageous timing of control operations. *Food lures* like methyl eugenol, sugar and propiononitrile, sinigrin and protein hydrolysate are other useful attractants used in trapping insect pests. *Sequential sampling* is most important as it can be made use of when pest populations are very high or very low.

It utilizes knowledge of economic threshold to determine whether further sampling is required for taking up insect control programme or not which saves time and labour. Low density of pest species suggests no action is needed and extremely high density warrants pest control action. For moderate pest density more sampling is required before taking up integrated pest management programme. (South Wood 1966; Ruesink and Kogan 1975).

The conditions which may influence the sampling results are lack of uniformity between individuals in their sampling techniques; time of the day at which sample is made; and size and stage of insect pest. These shortcomings can be overcome by perfecting sampling techniques.

Samplings will help in determining workable economic threshold against key pests. Once the pest densities are determined, a decision must be made whether or not to apply control measures. Insecticide should be applied only when damage or number of insects per given area reaches specific level. The level at which economic damage is caused by the insect pest is known as *economic injury level.* At this is population level loss in economic value of the crop due to insect damage is greater than the cost of the insecticide and labour required to control the insect pest.

It is necessary before taking up an integrated pest management programme against any key pest in a given agro-ecosystem to have basic information on its abundance and distribution, biological and behavioural characteristics, type of food, natural enemies by conducting large scale surveys and adopting appropriate sampling techniques. It is also desirable to know the effective control methods. This knowledge will help in determining the vulnerable stage of the key pest, the appropriate time of action and suitable effective methods which can be used at particular time for particular stage.

Components of Integrated Pest Management: They are systematically listed by Luckman and Metcalf (1975) and are as follows:

1. *Cultural Methods:* These methods involve manipulation of agronomic practices like variation in time of planting or harvesting; crop rotation, tillage of soil, pruning and thinning, fertilization, sanitation, water management, planting of trap crops, crop residue destruction and burning of stubbles. These practices help in reducing the pest populations.
2. *Physical Methods:* These methods include utilization of heat, cold, humidity, sound and energy in the form of light traps and light regulation.
3. *Mechanical Methods:* Such types of methods involve hand destruction of insect or its different stages of life cycle; exclusion by screens and barriers; trapping through suction devices and collecting machines and crushing and grinding.
4. *Biological Methods:* These methods include conservation of natural enemies, introduction, artificial increase and colonization of specific parasitoids and predators and use of specific bacteria, virus, protozoan diseases and fungus.

A good management programme must consider the introduction of new natural enemies of insect pest as well as conserve and augment already existing natural enemies.

5. *Chemical Methods:* These comprise insecticides, attractants, repellents, sterilants and growth inhibitors.

The strategy of a good integrated pest management programme advocates need based use of insecticides rather than calender based prophylactic treatments. An insecticide should be chosen on the basis of its effectiveness and minimum effects on useful insects and non-targeted organisms. Insecticidal formulations, application methods and timing of application are also important for effective integrated control programme. One such example has been demonstrated by Prem Kishore 1984.

Extensive insecticide evaluation programmes have led to the identification of insecticides that are effective against major insect pests. A wide range of selective wide spectrum commercial insecticides has been identified for important pests *viz.*, endosulfan, dimethoate, malathoion etc. An insecticide should be chosen on the basis of its effectiveness.

The application method is also an important aspect of the control programme. For example, soaking rice seedlings roots in an insecticidal solution or making a root coat treatment is cheaper than traditional methods and efficient in controlling early season pests.

Root zone application in rice of a systemic insecticide is also a very promising technique in several countries. This technique is less hazardous to parasites and predators and far superior to broadcast application or foliar spray.

Finally, timing up insecticide application to the most critical stage of the pest's life cycle or population build up is more effective than several calender based applications. This is borne out in results from experiments on brown planthopper and stemborers.

6. *Varietal Resistance:* The second component in a pest management programme is varietal resistance. Resistant rice varieties have been developed for almost all major insect pests and several such varieties are currently used in various Asian countries. Some varieties possessing resistance to several major pests have also been successfully developed. For example IR 40 has resistance to stemborers, green leafhopper, brown planthopper and whorl magot in rice, Two sorghusm varieties, *P 311* and *SPV 1015* (PGS 1) showed resistance to shootfly, stemborer, grey weevil and Pyrilla.

Judicious integration of chemical control and varietal resistance is a practical technique for improving a control programme. Because resistant varieties need fewer insecticide applications, the costs of insecticide treatment are reduced and the use of less insecticide favours the survival of natural enemies.

Although resistant varieties contribute substantially to a pest management programme, their stability against the possible development of insect strains to which they are not resistant must be constantly monitored.

7. *Genetic Methods:* They include release of genetically incompatible or sterile pests for suppression of population of insect pests.
8. *Regulatory Methods:* These include quarantine suppression and eradication programmes.

The use of resistant varieties of available and useful cultural practices alongwith limited use of a selective insecticide based on the economic threshold will add in conserving natural enemies and avoid disruption of environment. Together these practices will determine the success of any integrated pest management programme as these efforts aim to reduce key pest's equilibrium position permanently to a level lower than the economic threshold.

Utility of Integrated Pest Management

In a country like India evolving and implementing integrated pest control practices will enhance national economy as being more efficient and cheaper methods of control. Adopting integrated pest management strategy will avoid upsetting of balànce in nature as well as delay the development of pesticide resistant races of pests. This strategy will further ensure need based application of pesticides leading to deposition of considerably reduced quantities of harmful residues. Pesticide industry will also get a boost as the apprehensions that the pesticides have no role to play in pest management system will be set to rest.

The pesticides are definitely going to be important components of this system especially during pest epidemics.

The researches on integrated pest control have been in progress for the last three decades, but such efforts have been limited or rather fragmented in practice. However, they have certainly made their impact in crops like sorghum, maize, millets, cotton, sugarcane (Prem Kishore 1983, Prem Kishore and Vishwanath 1983, Aggarwal *et al.)* 1983, Gupta 1983, Prem Kishore and Gahukar (1992), fruits like apple, pear, soybean and vegetables (Flint and van den Bosch 1981).

Evaluation and Promotion

The evaluation and promotion of this concept in recent years has developed an awareness of certain essential features of the education and training for pest management. The success of integrated control practices will depend on a wide range of appropriate expertise on research/demonstration/training projects. It requires a strong research base, to provide information on insect pests, the crops, agronomic practices related to the crop and their interactions.

Once the various integrated control programmes are developed, it is necessary that they should be put to practice. The implementation of these programmes will require scientists/trained technicians and efficient extension media.

The scientists/technicians should have the knowledge and experience to: (i) recognise both beneficial and harmful insects; (ii) identify insect injury on plants; (iii) distinguish between insect injury plant diseases and other disorders created implants by abiotic factors; (iv) measure various environmental factors and their interactions; and (v) utilise various control measures. In addition to these, scientists/technicians should also possess communication skills which are needed to work with the farmers. In this direction National Institutes and various Agricultural Universities can play an important role by holding short-term or long-term training courses and carrying out problem oriented research. As it is multidisciplinary approach, professional scientists are required to collaborate in its planning and execution.

8

PEST CONTROL

The Entomology consist from "Greek" words. The means of "Entomo" is insect and means of "Logos" is knowledge.

- Entomology is the branch of zoology that deals with the study of insects under class insecta of phylum "Arthropods".
- The lac insect *Laccifer lacca* Kerr belongs to the order Hemiptera.
- The silk worms belong to the insect order Lepidoptera.
- Honeybees belong to the order of "Hymenoptera".
- It has been estimated that 50-70% of the grain crops are pollinated by insects, a major portion of it by bees alone.
- The latest production of honey in India is estimated at about 25000-30000 tonnes, its major portion obtained from "Apis *dorsata*".
- The all insects deficit RBC in blood.
- The blood colour of insects is green with yellow.

- The latest estimated total annual losses caused by insect pests to major crops and the food-grains in storage in India to the Rs. 336.6 billion.
- The world wide pre-harvest and post-harvest food losses due to pests have been estimated to be about 48%.
- There is a higher percentage of 46% loss for rice and a lower percentage of 24% for wheat.
- The higher loses occur in Africa 42% and Asia 43% respectively of their potential crop value.
- By the controlling insect-pests of sugarcane the yield could be increased by 20% from 229.2 to 284.9 million tonnes leading to an additional income of 43730 millions of rupees.
- Plant genetic system, a Belgium Biotechnology company in July 1987 were the first reported the development of transgenic plants of tobacco.
- Over 2.5 billion kg of pesticides are applied in the world annually.

Table: Pesticides Consumption durint 1999-2000 in India

Insecticides	60%
Fungicides	21%
Herbicides	14%
Others	5%

- Punjab have the major consumption technical grade pesticide in India, approximately 7150 metric tonnes.
- Lowest consumption of pesticide in India is Daman and Diu, approximately 1 metric tonnes.
- Bartlett (1956) coined the term "Integrated Pest Control"

The term "pest management" was given by Geier (1970).

Organised Pest Control

IPM as a dynamic and constantly evolving approach to crop protection in which all the suitable management tactics and available surveillance and forecasting information are utilized to develop a holistic management programme as part of a sustainable crop production technology (By Dhaliwal and Arora 2001)

- Destructive Insect and Pest Act was enforced in the year of 1971.
- The maximum percentage of insecticide is used in cotton in India.
- Average damage of food grains during stored period in India is 10%.
- Plant Protection and Quarantine Act was passed in the year of 1914.
- First insecticide discovered was D.D.T.
- The maximum amount of sugar in honey is 73%.
- 68% resin found in lac insect. *(U.P. Lecturship 2003).*
- Central Honey Bee Research Station is situated at Pune (Maharashtra).
- Lac Research Station is situated at (Bihar).
- Central Silk Research Station is situated at West Bengal.
- D.D.T. was first of all made by "Zedler" scientist.
- The first synthetic pyrethroid was described in 1973.
- LD_{50}: Lethal dose required for killing 50% population of test animals is known as LD_{50} value. The unit of LD_{50} is mg/kg of body weight of test animals or Hg/g body weight of insect.
- Biology has two subdivisions. BOTANY-which deals with plants and ZOOLOGY-which deals with animals.
- The aim of zoology is to collect knowledge of animals.

Disciplines in Zoology

Entomology	—	Study of Insects
Protozoology	—	Study of Protozoa
Nematology	—	Study of Nematodes
Rodentology	—	Study of Rodents
Ornithology	—	Study of Birds
Mammology	—	Study of Mammals
Ichthyology	—	Study of Fishes
Helminthology	—	Study of Worms
Acarology	—	Study of Mites

Diciplines in Entomology

Insect Taxonomy

Insect Morphology

Insect Anatomy

Insect Histology

Insect Physiology

Insect Toxicology

Insect Pathology Insect Parasitology

Insect Genetics

Applied Entomology

Agricultural Entomology

Medical Entomology

Horticultural Entomology

Veterinary Entomology

Storage Entomology

Forest Entomology

Industrial Entomology

Insects: Structure and Life

The Rasping-sucking mouth parts are intermediate in structure between the piercing-sucking and chewing type and are found in Thrips.

The Chewing-lapping type of mouthparts are found in honeybees.

The Piercing-sucking type of mouthparts are found in bugs and mosquitoes.

The Sponging-type of mouth parts are common in flies the typical example being that of the housefly.

- Pest: A pest is an organism whose population often increase above a certain level of economic injury and its existance conflicts with man's welfare, convenience and profits.

Sorts of Pests: There are three categories of pests

(i) Regular Pests: These are generally found in abundance during a crop season e.g. aphids, Jassids, and Thrips.

(ii) Sporadic Pests: These assume pest status occasionally in certain years and include locusts, grasshoppers, hairy caterpillars, crickets and cutworms.

(iii) Potential Pests: These pests normally cause negligible damage but may become highly destructive resulting from some disturbance in the environment and the consequent increase in their number e.g., armyworm on wheat.

Measures for Insect Management

(a) Legislative Measures: Legislative control involves enactment of laws to regulate the entry, establishment and spread of pests.

Legislation is also imperative to stop the accidental entry, from outside the country of certain pests which may not be present in that country.

Thus legislation is of four kinds

(i) Legislation for foreign quarantine to prevent the introduction of new pests from abroad.

(ii) Legislation for domestic quarantine to prevent the spread of established pests within the country or within a particular state.

(iii) Legislation for notified campaign of control against pests.

(iv) Legislation to prevent the adulteration and mishandling of insecticides or other devices used for the control of pests.

(b) Cultural Practices: The cultural methods of insect control comprise regular form operations, which are so performed as to destroy the insects or to prevent them from causing injury.

For achieving cultural control, the various agricultural practices can be categorised under various groups

1. Tillage
2. Clean seed
3. Irrigation
4. Fertilizers
5. Clean culture
6. Crop rotation
7. Trap crops
8. Pruning and Thining
9. Intercropping
10. Destruction of crop residues.

(c) Mechanical and Physical Control: Mechanical and physical control measures involve the use of force or physical factors of the environment with or without the aid of special equipment.

Manual Labour

It means working with hands. Sometimes with the aid of some simple equipment, like bags, nets etc.

(d) Chemical Control: The use of chemicals for the control of insect-pests. The classification of insecticides is as follows.

Categories of Insecticides

Insecticides can be grouped in various ways:

According to the Mode of their Entry

(1) Stomach Poisons: They are generally used against insects with chewing type of mouthparts and under

certain conditions against those with sponging, siphoning, lapping or sucking mouth parts.

(2) Systemic Poisons: A systemic insecticides when applied to seeds, roots, stems or leaves of plants is absorbed and translocated to various parts of the plant in amounts lethal to insects which feed on them, like-dimethoate, phosphamidon, phorate, aldicarb etc.

(3) Contact Poisons: The contact poisons are applied as sprays or dusts either directly onto the body of insects or to the places frequented by them. These poisons kill the insects either by clogging spiracles and respiratory system or by entering through the cuticle into the blood and acting as nerve or general tissue poisons. like-chlorinated hydrocarbons (DDT, HCH, aldrin etc.) carbamates (carbaryl, aldicarb, carbofuran etc.) and organophosphates (Malathion, Monocrotophos etc.)

(4) Fumigants: Poisonous gases, derived from either solids or liquids are used as fumigants to kill insect pests of stored grains and other products in warehouses, museum, godowns etc.

- The common fumigants are-sulphurdioxide, phosphine, methylbromide, EDB etc.

According to Mode of Action

Physiological poisons

Protoplasmic poisons

Respiratory poisons

Nerve Poisons

According to their Chemical Composition

(i) Elements: Such as sulphur, phosphorus, thallium, mercury etc.

(ii) Inorganic Compounds: Such as lead arsenate, sodium fluosilicate, zinc phosphide, paris green etc.

(iii) Organic compounds

(a) Compounds of plant origin such as pyrethrum, nicotine, rotenone etc.

(b) Animal and mineral oils such as fish oil, diesel oil etc.

(c) Synthetic organic compounds such as DDT, Malathion, carbaryl etc.

(iv) Poisonous Gases: Such as hydrogen cyanide ethylene dichloride, carbon tetrachloride, methyl bromide, phosphine etc. used as fumigants.

(a) Nematicides: These chemicals are used for controlling plant parasitic nematodes. Some of common nematicides are mentioned here.

(1) DBCP (2) D-D Mixture

(3) Ethylene dibromide (4) Ethoprop

(5) Phorate (6) Aldicarb

(b) Rodenticides: These chemicals are used for control of rodents such as - strychnine, warfarin, zinc phosphide etc.

(c) Molluscicides: Molluscisides may be classified into two broad categories.

(i) Aquatic Molluscicides: copper sulphate niclosamide, sodium pentachlorophenate etc.

(ii) Terrestrial molluscicides: include carbamates-aminocarb, methiocarb, mexacarbate etc.

(d) Fungicides: These chemicals used against fungus like

(i) Bordeaux mixture

(ii) Sulphur and allied compounds

(iii) Carbamates

(iv) Organic mercury compounds

(v) Antibiotics

(vi) Nitrogen compounds Pesticide Application Equipment

The various equipment used include dusters, sprayers, vapour and smoke generators, agricultural aircrafts, granule applicators etc.

The Insecticide Dusters: The insecticides dust which are very fine particles when falling free in air either slowly or driff for a long distance due to wind.

- Parts of the Dusters

(i) Hopper or Container: It is used for holding the dust.

(ii) Blower: It work to throw the dust outside from container to the discharge outlet.

(iii) Handle: It is used to rotate the gear and there by fan throws the air from blower.

(iv) Feeding Brush: It agitates the dust in the hopper to prevent uneven dusting.

(v) Feed Mechanism: It control the rate of powder coming from hoppers.

(vi) Discharge Line: It carries the powder from bowler to nozzle.

(vii) Shoulder Streps: Shoulder streps are provided with mounting frame in some dusters to set whole mechanism on the chest.

Sorts of Dusters

(a) Manually Operated Dusters
   (i) Plunger duster
   (ii) Bellows dusters
   (iii) Rotary or fan dusters
   (iv) Wet dusting equipment

(b) Power Operated Dusters
   (i) Tractor mounted duster
   (ii) Engine-operated duster

Merits of Dusters

1. Large area can bc covered
2. Less expensive
3. Easy to operate
4. Very effective in arid zone area Demerits
5. Loss of chemicals due to drift problem

6. Less efficient
7. Increase toxic hazzard
8. Sprayers

A sprayer is an appliance which atomizes the spray fluid, which may be a suspension, an emulsion or a solution, the fluid is ejected with some force for proper distribution.

The various types of sprayers can be grouped into four categories, depending upon the volume of spray fluid discharge to cover a unit area.

1. High Volume Sprayer: These require 300-500 litres per ha. of spray fluid. For example-knapsack sprayer, stirrup sprayer, rocker sprayer, compression sprayer, foot sprayer etc.
2. Low Volume Sprayer: These sprayers require 50-100 litres of spray fluid per ha. for instance-motorized knapsack sprayers.
3. Ultra Low Volume: These sprayers require 1-5 litres per ha. of spray fluid. For example knapsack mist blowen fogair.
4. Aerosols: Less than 1.0 litres per ha. of spray fluid is required to create fog. The example are-pressurised containers, swing fog machine.

Kinds of Sprayers

(a) Manually Operated Sprayers

These are of two types

(i) Pneumatic sprayers
   (a) Hand compression sprayers
   (b) Pressure retaining pumps.
   (c) Small pneumatic atomizers.

(ii) Hydraulic sprayers
   (a) Knapsack sprayer (b) Foot sprayer
   (c) Rocker sprayer (d) Stirrup sprayer

(b) Power Operated Sprayers

(i) Mist blower or motorized knapsack sprayer

(ii) Tractor mounted sprayer

(iii) Power sprayer

(iv) Ultra-low-volume sprayer

(v) Aerosol dispensers

(vi) Smoke generator

(vii) Vapour generators.

(c) Agricultural Aircrafts

(i) Light aircraft

(ii) Medium aircraft

(iii) Heavy aircraft

(d) Granule Applicator

Placement of insecticidal granules in the whorls of crops like-maize, jawar, etc., is very effective for the control of stem borers.

(e) Soil Injectors

Most soil fumigants are liquids and are applied with soil injectors.

- Merits of sprayers

1. Less drift problems as compared to dusters.
2. It is handy and simple to use.
3. Used to apply various types of chemical such as bactricide, fungicide, insecticide etc.
4. Most commonly used appliances.
5. Covers more area in less time.
6. Treatment can be done on height and also blow the leaves.

Sustainable pest management is otherwise known as IPM i.e., Integration of tactics of control a single pest on one or more crops. IPM as applied in agriculture is ideally the use of most effective, economical, safest, ecologically sustainable and sociologically acceptable combination of physical, chemical and biological methods to limit the harmful effects of crop pests.

According to FAO panel of experts (1966-72) IPM is-"A pest management system that in the context of the associated environment and the population dynamics of the pest species,

utilizes all suitable a manner as possible and maintains the pest population at levels below those causing economic injury."

The overall objective of IPM is to create and to maintain situations in which insects are prevented from causing significant damage to crops.

A number of tools or components have been successfully used for IPM in several crops. These include-use of pest resistant or tolerant, predators and pathogens, use of parasites, summer ploughing, quarantine measures, late planting, hard collection and destruction, judicious use of pesticides, attractants, repellents, sterilants, growth regulators.

- The sustainable pest management in India has yet to come to be well recognized, but it is still more of an aspiration than a reality for the average farmers.

Biopesticides in Significance

Biopesticides are most important, effective and commercially viable as these are inexpensive cause no pollution. Pose no risk to human health and have a long term capability to control pests. Biopesticide agents are abundantly available in nature. Several pathogens including viruses such as-NPV (Nuclear polyhedrosis viruses and granulosiss viruses (G.V.)

- Bacteria like - *Bacillus thuringiensis*.
- Fungi like-Metarhizium, Beauveria, Verticillium.
- Protozoa like- *Schizogregrine* cause diseases in insects and destroy them. Similarly, several insect-parasitoids (Parasites thriving on insects are also known in nature.
- *Trichoderma* is an egg parasitoid of several pests.

	Common Name	Scientific Name	Order	Host Plants	Damaging Stage	Mouth Parts
1.	Locust	*Schistocera gregaria* Forsk	Orthoptera	Polyphagous adult	Nymph and cutting type	Chewing and
2.	Termite	*Odontotermes obesus* Ramb	Isoptera	Polyphagous	Nymph and adult	Chewing and cutting type
3.	White Grub	*Holotrichia consanguinea*	Coleoptera	Polyphagous	Grub and adult	Biting and cutting type
4.	White Fly	*Alerolobus baradensis* Mask	Hemiptera Homoptera	Sugarcane, Jawar, Bajra, Wheat, Barley	Nymph and adult	Piercing and sucking type
5.	Leaf Hopper or Pyrilla	*Pyrilla perpusilla* Walk	Hemiptera Homoptera	-do-	-do-	-do-
6.	Sugarcane top borer	*Tryporyza nivella*	Lepidoptera	Sugarcane, Jawar, Bajra, Kans, Sarkanda	Larva	Biting and cutting type
7.	Shoot borer	*Chilo infuscatellus* Snell	Lipidoptera Jawar, Maize	Sugarcane,	Larva	Biting and cutting type

Contd....

	Common Name	***Scientific Name***	***Order***	***Host Plants***	***Damaging Stage***	***Mouth Parts***
8.	Paddy stem borer	*Tryporyza incertulas*	-do-	Paddy	-do-	-do-
9.	Rice gundhi bug	*leptocorisa varicornis*	Hemiptera	Paddy, Jawar, Bajra Maize	Nymph and adult	Piercing and sucking type
10.	Paddy plant hopper	*Nephotettix* spp.	Hemiptera	Paddy	-do-	-do-
11.	Cotton Jassid	*Amrasca bigutulla bigutulla*	Hemiptera	Cotton, bhindi	Nymph and adult	-do-
12.	Cotton leaf roller	*Sylepta derogata*	Lepidoptera	-do-	Larva	Chewing and cutting type
13.	Spotted boll worm	*Earias vitella*	-do-	-do-	-do-	-do-
14.	Pink boll worm	*Pectinophora gossypiella* Sound	Lepidoptera	-do-	-do-	-do-
15.	Mustard aphid	*Lipaphis erysimi*	Hemiptera	Radish, Rai, Mustard, Cauliflower, turnip etc.	Nymph and adult	Piercing and sucking type

Contd....

Common Name	Scientific Name	Order	Host Plants	Damaging Stage	Mouth Parts
16. Painted bug	*Bagrada cruciferorum*	Hemiptera	-do-	-do-	-do-
17. Cut worm	*Agrotis ypsilon*	Lepidoptera	Mainly gram and potato	Larva	Cutting and biting type
18. Gram caterpillar	*Helicoverpa armigera*	-do- tomato, Arhar	Gram, pea,	-do-	-do-
19. Red pumpkin beetle	*Raphidopalpa foveicollis*	Coleoptera	Plants of cucurbitaceae	Grub and adult	-do-
20. Fruit fly	*Dacus cucurbitae*	Diptera	Plants of cucurbitaceae	Maggot	-do-
21. Epilachna beetle	*Epilachna vigintioctapun-ctata*	Coleoptera	Brinjal and cucurbits	Grub and adult	-do-
22. Brinjal shoot and fruit borer	*Leucinodes arbonatis*	Lepidoptera Potato	Brinjal and	Larva	-do-
23. Diamond back moth	*Plutella maculipensis*	Lepidoptera cruciferae	Plants of biting type	-do-	Cutting and

Contd....

	Common Name	Scientific Name	Order	Host Plants	Damaging Stage	Mouth Parts
24.	Tobacco caterpillar	*Prodenia litura*	Lepidoptera	Tobacco and cruciferae	-do-	-do-
25.	Mango hopper	leaf *Amritkodus atkinsoni*	Hemiptera Homoptera	Mango adult	Nymph and sucking type	Piercing and
26.	Mango mealy bug	*Drosicha mangifera green*	Hemiptera Homoptera	Mango, Jack fruit etc.	-do-	-do-
27.	Lemon buterfl	*Popilio demoleus* Linn	Lepidoptera	Citrus	Larva	Biting and cutting type
28.	Ber fruit fly	*Carpomyia vesuviana*	Diptera	Ber	Larva	-do-
29.	Rice weevil	*Sitophilus oryzae* Linn.	Coleoptera	Stored grain	Grub and adult	Chewing and cutting type
30.	Khapra beetle	*Trogodemua granarium*	Coleoptera	Stored grain	Grub	-do-
31.	Grain and flour moth	*Siiotroga cerealella.*	Lepidoptera	-do-	Larva	-do-
32.	Pulse beetle	*Callosobruchus chinensis*	Coleoptera	Pulses	Grub and adult	-do-

Contd....

Common Name	*Scientific Name*	*Order*	*Host Plants*	*Damaging Stage*	*Mouth Parts*
33. Lesser grain borer	*Rhizopertha dominica*	Coleoptera	Stored grain	-do-	-do-
34. Red hairy caterpiller	*Amsacta spp.*	Lepidoptera	Polyphagous	Larva	Biting and cutting type
35. Gurdaspur borer	*Acigona steniellus*	Lepidoptera	Maize, Jawar, Sugarcane	Larva	-do-
36. Army worm	*Mythimna separata*	Lepidoptera	Paddy, Jawar, Bajra, Wheat, Maize, Sugar.	Larva	Chewing and cutting type
37. Shoot fly	*Atherigona varia soceata*	Diptera	Jawar, Bajra .	Maggot	-do-
38. Fruit sucking moth	*Ophideres* spp.	Lepidoptera	Mango, Guava, Citrus and Pomegranate	Moth	Siphoning type
39. Potato tuber moth	*Gnorinoschema operculella*	Lepidoptera Tomato, Tobacco and *Agricultural Entrance*	Potato,	Larva 175	Chewing and cutting type Brinjal

Important Glossary

- Acaricide or Miticide: The chemical substances used to kill mites are known as acaricide.
- Adulticide: Chemical substance used to kill adult insects.
- Antibiosis: Antibiosis refers to the adverse effect of the host plant on the biology of the insects and their progeny infesting it.
- Allelochemicals: Allelochemicals are non-nutritional chemicals produced by an organism of one species and affect the growth, health, behaviour or population biology of individuals of another species.
- American Bolworm: Recently the American bollworm has become a serious pest of cotton in northern parts of India. It was first reported on cotton in 1977 in Punjab.
- Armyworm: The armyworm is a pest of graminaceous crops-all over the world like jawar, maize, bajra etc.
- Antifeedants: The antifeedants are chemicals which inhibit or deter the feeding of insects due to their presence on the natural food of the species concerned. An antifeedant acts by suppressing the gustatory receptors.
- Avermectins: Avermectins are macrocyclic lactones which were originally isolated in 1976 by scientists at Merck and Co, Inc from a culture of *Streptomyces avermitilis* from Japan.
- Allomone: A compound released by one organism which evokes a reaction in an individual of a various species that is favourable to the emitter but not to the receiver.
- Antimone: A substance produced or acquired by an organism that when it contacts an individual of another species in the natural context, evokes in the receiver a behavioural or physiological reaction that is maladaptive to both the emitter and the receiver.

9

New Technology in Plant Protection

Although agricultural engineering is a fairly new subject, its effect is being felt in India increasingly. In recent years it has contributed considerably to agricultural production in the country both directly and indirectly. The number of modern machines used in agriculture is a fairly good index of the progress made during the four Five-Year Plans. Roughly, more than 1,00,000 tractors, crawler-tractors and power-tillers are in use in India.

Similarly, the number of combines, self-propelled, trailed and mounted ones, has gone up to more than 1,500. The number of power-threshers may be in the vicinity of 5,00,000 and the number of both diesel and electric pumps must have crossed the million mark. And yet all this progress mostly made during the last 25 years is insignificant as compared with the use of modern agricultural machines in the developed countries, such as the USSR, the USA, Japan, the UK, Canada and Australia, and we have yet to cross a number of hurdles, but the progress made so far is quite encouraging and forms a basis for a leap forward. This development has, however, made two things clear:

(1) that the Indian farmers, though belonging to a less developed country, are quick in adopting new methods, techniques and machines, provided they are convinced of their utility, and

(2) that the Indian industries can manufacture as good modern machines as those manufactured elsewhere. The confidence created by these developments augurs well for taking bold steps in the future. And this in itself is no mean achievement.

Agricultural engineering, as defined by international institutions, comprises four main branches, namely:

(i) farm implements and power,

(ii) rural structures,

(iii) soil conservation, drainage and irrigation, and

(iv) rural electricity.

Because of the peculiar local conditions, the lack of development of electricity on a large scale, and the immediate utility of implements, agricultural engineering in India has come to mean more of agricultural implements and machinery. But, in fact, agricultural engineering is a very vast subject. And may be defined as the application of the knowledge, techniques and disciplines of various fields of engineering to the solution of problems arising in the fields of agriculture and rural living, with the object of reducing labour, improving agricultural productivity per worker, and raising the standard of living of the farmers and increasing the overall earnings per worker.

During the Second and the Third Five-Year Plans, agricultural engineering divisions were added to the departments of agriculture in the states and at the Centre. They undertook the work of hiring or selling tractors, oilengines, implements, pumps, etc. Several training institutions and workshops were started to train artisans and mechanics, and in every state a research cum testing and training centre for agricultural implements was established. The cumulative effect of all these steps has been to popularize mechanization and to prepare a sound base for it.

Then came an era of more intensive planning and reorganization on a wider scale. With the establishment of agricultural universities, several agricultural engineering colleges were established, offering graduate and post-graduate courses. In most of the states, where such colleges were established, they took over research, education and extension from the state departments of agriculture. The most important step in the reorganization has been the establishment of agro-industries corporations, one in each state, with the financial assistance of the Central Ministry of Agriculture. The main objectives of establishing these corporations have been to take over the supply and service functions of the departments of agriculture and to expand them. These agro-industries corporations have each a paid-up capital of 2 to 5 crores of rupees. They have taken over the existing government workshops also. The progress made by these agro-industries corporations is given in this chapter.

The other most important development has been in the field of industry which has established so far half a dozen factories for manufacturing tractors, 3 or 4 for manufacturing power-tillers and several other factories for manufacturing oil-engines and electrically-driven pumping-sets. Some of these industries are being substantially assisted by the Industrial Finance Corporation of India. The Government has also established a factory for manufacturing crawler tractors and another one for manufacturing wheel-type tractors. These three main developments helped to quicken the pace of mechanization of Indian agriculture.

The following tabulated statement clearly shows the progress made during the last 25 years:

Sl. No	Name of the Implements	1951	1956	1961
1.	Mouldboard-type iron ploughs	9,30,000	13,66,000	23,00,000
2.	Bullock-driven sugarcane crushers	5,04,000	5,44,000	5,90,000
3.	Power-driven sugarcane crushers	21,000	23,000	32,210
4.	Centrifugal pumps with oil-engines	82,000	1,22,000	2,29,000
5.	Electric pumps	25,000	55,000	1,60,000
6.	Tractors	8,500	21,000	34,000

It is not merely the knowledge of agricultural engineering, including implements and machinery, which is necessary for increasing agricultural production, but also the knowledge of the sources of their availability, prices and upkeep. The following is a list of sources through which the information on the availability and prices of the machinery can be obtained.

(1) At the Village Level: the Village-Level Worker or *gram sewak.*

(2) At the Block Level: the Agricultural Extension Officer or the District Planning Officer.

(3) At the District Level: the District Agricultural Officer or the District Planning Officer.

(4) At the State Level: the State Agricultural Engineering Section and the Managing Director of the State Agro-Industries Corporation.

(5) At the Central Government Level: The Director of Agricultural Implements and Joint Commissioner (Machinery), Ministry of Agriculture and Irrigation, Krishi Bhavan, New Delhi. The private industry as well as the government, through their agro-industries

corporations, are in the field now for the supply of agricultural machinery to the farmers. To prevent the charging of exorbitant prices for tractors, the government has fixed the maximum prices of tractors with the assistance of the Bureau of Cost Accounting in the Ministry of Finance. This organisation has helped the farmers very much.

Future Prospect: Agricultural aviation is now gaining ground in India. At present, the aeroplanes and helicopters are being used mostly for spraying pesticides, but in future, they may be used for emergency sowing, and spraying liquid fertilizers on the growing crops. Agricultural electronics has also begun and the remote-control tractors are being designed and tested in some countries. Japan has successfully designed machines for transplanting paddy seedlings and their trials are going on at the Tractor-Training and Testing Centre at Budni. A very ambitious programme for joining the Ganga with the Kauveri, via the Mahanadi, the Narmada, the Godavari and the Krishna, is under the consideration of the Central Government and some preliminary work is being carried on. With the successful explosion of an atomic device, India may be in a position to generate more electricity in future. The availability of plentiful irrigation water as a result of the Ganga-Kauveri Project, and the availability of more electricity from atomic energy will make the progress of farm mechanization in India speedier. Thus the future seems to be bright.

Role of Machinery in Agriculture

The beginning of mechanization of Indian agriculture was made by the use of improved hand-tools and improved bullock-drawn implements. Particular stress was laid on these items during the first three Five-Year Plans and slowly more sophisticated implements were introduced during the Fourth and Fifth Plans. These included power driven threshers, tractors and the combine harvesters. Since simpler implements were taken up first and since in many areas they are still being used, their description is taken up here first. Quite a variety of implements and machinery is required for carrying out various agricultural operations, starting from the opening of the land to harvesting, threshing, winnowing and storage. These operations include ploughing, harrowing, levelling, sowing, intercultivation, the application of fertilizers and manures, harvesting, threshing and winnowing.

India is a sub-continent and a variety of indigenous or *desi* implements are in use in different parts of the country for the last so many centuries. Before efforts were made to introduce foreign implements, it was necessary that a survey of indigenous agricultural implements used in India should be undertaken. Therefore, under the Indian Council of Agricultural Research, a country-wide survey was carried out and its report was published. This survey has brought a number of important facts to light. The number of implements used by the Indian farmers is really very large. Some of these implements, even if they look crude, are really useful and the consensus of the agricultural engineers in India has been that it would be easier at least as a first step to improve upon the indigenous implements in order to increase their efficiency and thus give to the Indian farmers something which is within their means.

To achieve this objective effectively, the Indian Council of Agricultural Research has now established one research-cum-testing and training centre in each state. Some of them have now been converted into regional centres or have been taken over by the state governments. The object is to improve the indigenous agricultural implements, to design new implements and to test them in the field up to the prototype level. After the implements have proved to be useful, the prototypes can be given to the manufacturers both in the public as well as in the private sectors and the implements can then be sold through them to the farmers.

Research Principles: In India, the holdings are generally small, the average farmer is poor, and the draft cattle are also small and ill-fed. Taking these factors into consideration, certain principles, on which research is to be carried out, have been laid down. These are given below:

(1) That the implements and machines for the Indian farmers should be simple in construction so that they could be operated by the illiterate farmers, and should either be manufactured or could be repaired by the village artisans or mechanics.

(2) The prices of the implements should be within the reach of the ordinary farmers, though they need not necessarily be cheap at the cost of efficiency, but they should also not be too costly. In the case of power-driven equipment, the difficulty about the cost can be overcome by giving the machinery on hire or hire-purchase system or on custom basis or can be utilized on co-operative basis.

(3) The implements should be light, so that they can be transported from the village to the field and that the draft of the implement is suitable for the types of animals that are used in particular areas.

(4) That the implements should be scientifically tested on a fairly uniform basis throughout India, so that the results are known and can be compared. There should be as little guess work as possible.

(5) In designing the implements, as far as possible, the local available materials are to be used. In the case of soil-working and plant-working parts, durable materials, such as high-carbon steel, should be used.

To fulfil these requirements, the Indian Council of Agricultural Research has also finalized procedures for the testing and designing of implements and the proformas in which the test data are to be collected. In recent years, the attention of the Indian farmers has been drawn to the power-operated machines and equipments, such as tractors, oil-engines, and pumps. Quite a large number of the farmers are using these implements and firms have also been established in India to manufacture them. This trend towards mechanized farming is bound to increase.

Recently, however, new developments have taken place to bring about mechanization of agriculture. Owing to the shortage of diesel-oil and its increased price, the mechanical operation will have to be restricted to deep ploughing, land-levelling, clearing land, and other operations which cannot be carried out with bullock-power. Now greater attention has to be paid to the bullock drawn implements and manually operated implements. For instance, for short distances, bullock-carts have been found more economical than trucks; hence the necessity for improving bullock-carts and also bullock-drawn, but improved implements, on priority basis. The description of some improved implements is given in the following pages:

Primary Implement, Plough

A plough is the primary implement used all over India. The *desi* ploughs have almost a common design and they differ mostly in size, depending upon the type of the soil and the size of the bullocks or buffaloes used in the particular area. The heaviest ploughs are found in the black-cotton-soil areas of India. The *desi* plough is almost like a western-style cultivator, both in design and action. The body and the wedge may be of one piece or of two pieces. The iron point or share is made up of a rod or bar of mild steel, and penetrates into the soil and protects the wooden wedge against wear and tear. The beam varies from 2.5 to 3.5 m in length and is made of wood. The handle is fixed to the beam at the back of the body. All indigenous ploughs are almost similar in design, but vary considerably in size and weight to suit the prevailing soil types, ploughing conditions and the weights and heights of the draft animals in the locality. The working draft varies from 27 to 148 kg. The following table gives an idea of the size and the draft of the *desi* ploughs used in different states.

The *desi* plough makes a V-shaped furrow and, unlike a mouldboard plough does not turn over the soil. It is mostly made of hard wood, e.g. *babool* or *kikar* (*Acacia nilotica)*. The depth of ploughing is usually from 10 to 2.0 cm and the width from 13 to 18 cm. There are minor variations in design, weight, the length of the beam, etc. The duck-foot type of plough, as used in Orissa is different in design. In has a flat bottom of about 38x30 cm and is said to be useful under soil conditions in Orissa. The plough used in Maharashtra is very heavy requiring sometimes 2 to 4 pairs of bullocks. It is useful for ploughing the black cotton soils of the Deccan. Most of the *desi* ploughs are light in weight and can be carried either on the yoke of the bullocks, or on the shoulders by the farmers.

The depth of the plough can be adjusted with a wedge provided in the body of the plough or by adjusting the yoke either forward or backward on the beam. If the yoke is fixed forward, the plough goes deeper and if the yoke is taken backward, the plough goes shallower. In Andhra Pradesh, a simple triangular plate is fixed in front of the body of the plough so that ridges and furrows can be made. This is an ingenious device to convert the *desi* plough into a ridger. Since the *desi* plough acts like a cultivator, repeated ploughings and cross ploughings are necessary to secure a good tilth. The *desi* ploughs are also used for making shallow furrows for depositing seeds, for inter cultivation between the rows of crops and to harvest crops such as potatoes, groundnuts and sweet-potatoes. It can thus be seen that the *desi* plough is a multipurpose implement difficult to be replaced, unless more efficient multipurpose improved implements are designed.

Weight, Draft and the Furrow Size of Different Desi Ploughs

Ploughs	*Weight (kg)*	*Length of bean (m)*	*Average draught (kg)*	*Horse power expended*	*Average size of furrow (cm)*	
					Width	*Depth*
Bareilly	10.88	2.66	28.57	0.40	17.78	10.16
Pusa (Bihar)	15.42	2.5	27.66	0.43	26.4	17.78

Contd...

Ploughs	*Weight (kg)*	*Length of bean (m)*	*Average draught (kg)*	*Horse power expended*	*Average size of furrow (cm)*	
					Width	*Depth*
Angul (Orissa)	10.88	2.15	47.17	0.73	27.94	15.42
Sabour (Bengal)	19.95	2.87	100.24	1.35	30.4	20.32
Patna	15.87	2.71	54.43	0.81	27.94	20.32
Kerala	23.58	3.32	134.26	1.87	30.4	20.32
Dohad (Mumbai)	13.15	2.84	158.75	2.67	22.86	22.86
Coimbatore	37	3.53	328.0	1.71	27.94	22.00
Indore	57	3.15	403.5	1.71	22.00	26.00

Techniques of Bladed Harrows

These are called *bakhars* or *guntakas. Bakhar* is a very common implement used in the Deccan. It is one of the best implements for primary tillage. The farmers use it for preparing the land and for sowing the seed. It consists of a steel blade of about 45 to 90 cm in length and 8 cm in width. It can cultivate the land to a depth of 9 to 10 cm. Even the dryland tillage is done with this implement by adding a wedge to it. If the soil is very dry, a man stands on the implement to give it the weight. It can cover 0.8 to 1.6 ha in a day because of its greater width. The Lanka *bakhar* used in the Vidarbha area has a blade even up to 1.5 m in length and is used after a few monsoon showers to kill young germinated weeds. The implement is worked cross-wise in the fields to achieve a better tilth. This is a very useful implement and costs only Rs 30 to 45. Somehow, this implement has not yet been introduced into areas above the Narmada river, except in some pockets near Jhansi and Gwalior. It needs to be examined whether such an implement can be introduced into Bihar, Uttar Pradesh and Punjab. Almost every farmer in the Deccan possesses this implement. It is interesting to mention here that in some parts of the USA, where moisture conservation is an important problem, the use of "sweeps" and "chisel cultivators" which are similar to the Indian *bakhar* or blade harrow in working, is being advocated. The Egyptian and West Asian ploughs are also very much like the Indian bladed harrow. Recently, improvements have been made in the design of the *bakhars,* making it possible to change the blade as desired. They are mostly made of steel.

Mouldboard an Improvement in Plough: A mouldboard plough consists of a share and a mouldboard of steel. This is probably one of the first implements to be introduced into India from foreign countries. It differs in action from the *desi* plough in ploughing a square furrow and in inverting the soil either completely or partially, depending on the curvature given to the mouldboard. For a proper functioning of the mouldboard ploughs, the provision of an adequate vertical and horizontal suction to the share up is necessary. In the case of long-beam ploughs, the beam and the handle are made of wood, whereas in the case of a short-beam plough, the complete implement is made of metal. Where it is necessary to have a one-way ploughing, the turnwrest type of ploughs are used. Lakhs of such ploughs are in use in Maharashtra.

There, the ploughs are now available on hire also. It is being adopted where green-manuring crops or surface vegetation have to be buried to hasten decomposition. Heavy mouldboard ploughs require sturdier animals, and sometimes two or three pairs of bullocks are used. They are also employed for the eradication of deep-rooted weeds, e.g. *kans (Saccharum spontaneum)* and *kunds (Cyperus rotundus).* The turnwrest type of mouldboard ploughs are very efficient. For lighter ploughing, the beam-type ploughs are used. The shares of the lighter ploughs should be made of steel instead of cast iron. These ploughs are finding increased favour in several parts of northern India and are replacing the *desi* plough. The mouldboard plough in common use is with bullock-power cut furrow slices varying from 12 to 23 cm in width. Typical figures in respect of their relative performances are given below:

Double Mould Boards: These ploughs have double mouldboards. There are three sizes—the light, the medium and the heavy. The heavy ones are used for the cultivation of sugarcane and the lighter ones for the cultivation of vegetables and other crops. Some of the improved ridger ploughs have an arrangement for adjusting the width of the furrow by changing the distance between the two mould boards. For vegetable orchards, and for farmers who grow sugarcane, potatoes, etc., the possession of the ridger is a must.

Width of the plough (cm)	-	17.78	16.31 (nominal)	10.43 (nominal)
Width of the furrow cut (cm)	-	17.78	13.34 to 16.31	9.16 to 10.43
Depth of the cut (cm)	-	12.7	12.7	12.7
Drought in dry sandy loam at Delhi (kg)	-	123.37	127	118
Drought per square inch of furrow cross-section (kg)	-	0.562	0.632	0.913
Speed of bullocks (km)		3	1.673	1.737

Harrows and Cultivators: The indigenous harrows and cultivators are known as *bakhars* and *guntakas.* They mostly consist of a long sharp blade fixed to a wooden implement. The body of such an implement is made of *babool* wood, cut into a rectangular or octagonal shape. The length of the body is 91 to 182 cm and the width is 25 to 30 cm. The blade is attached to the log on its underside with two slanting wooden pegs. These implements are used for preparing the seedbed, for covering the seed and also for summer cultivation, provided the soil is not too hard.

Improved *bakhars* and *guntakas* have also been, lately, manufactured. The improvements relate to the changing of the blades, the adjustment of the depth and the attachments that may be needed in the case of these harrows. The length of the blade is from 61 to 152 cm and, therefore, this implement can cover quite a large area in a day. Steel *bakhars* and *guntakas* are nowadays mass-manufactured in India. Their prices range from Rs 40 to 55.

Some of the harrows used in foreign countries, e.g. the spike-tooth harrow, peg-tooth harrow, the spring-tooth harrow, and disc harrows have also been introduced. Out of all these harrows, the disc harrows have proved to be very popular. The spike-tooth harrow is also often used, particularly in parts of Punjab and western Uttar Pradesh. The spike-tooth harrow is also often used for inter-cultivation and for thinning crops in Punjab. The triangular frame on which spikes are fixed is used in that state.

Role of Soil Implements

Soil for Scoop: A soil scoop consists of a shallow box-like structure with an open front. On its front, a sharp blade is fixed and two hooks are attached to it. The bullocks are yoked to the hooks. Two wooden handles are provided at the back. The capacity of the soil scoop is 0.056 to 0.084 cu m. This implement is used when the soil is to be carried to a fairly long distance. The price of the soil scoop is Rs 60 to 1,000.

Wheel Barrows: Common wheel barrows are very useful for gardens or orchards. The wheels may be of mild steel or of rubber. The price of a wheel barrow is Rs 150 to 200.

Keni Making or Levelling Karaha: It consists of a fiat blade 1 to 5 m long, with a vertical handle and two rings in the front for yoking the bullocks with ropes. This is probably the commonest implement used in soil conservation for making bunds and levelling the land. Literally, thousands of these implements are used in Ahmadnagar district of Maharashtra for bunding. The bunding of about 200,000 ha has been completed with this implement. It is also becoming popular in other parts of Maharashtra, Andhra Pradesh and Karnataka for bunding land as well as for bunding *nullahs.*

The space where the bund is to be formed is first marked, and on both sides of this space, a width of about 5 m is ploughed with a mouldboard plough. The loosened soil is then moved on to form a bund. If the land to be ploughed is hard and cloddy, it hurts the feet of the bullocks and reduces their life. As far as possible, the bunds should be formed when the soil is soft. The price of the *keni* is Rs 40 to 60.

Patela: The indigenous implement is called *patela* or *sohaga*. It is merely a flat log 1.8 to 2 m long and is dragged over the clods by bullocks. The driver usually stands on the log to put weight on it and hold it down.

In Uttar Pradesh, the improved *pata* or *patela* is now mass-manufactured and has become quite popular. It is called *Singh patela*. It is provided with a series of pointed hooks made of 10 mm to 15 mm rods of mild steel. These hooks help to crush the clods and to collect the weeds. This implement costs 100 to 150 rupees, depending upon the length of the wooden blade and the number of steel rods used.

A land-leveller or *jandra* is also used in some tracts. It consists of a plank of wood, about 1 to 1.5 m long, 12 cm and 2.5 cm thick. In those areas where bamboo is readily available, a ladder-shaped leveller made of bamboo is used.

Bunds Forming: For soil conservation, sometimes it is necessary to make temporary low-level bunds, 18 to 22 cm high. These bunds last for one agricultural season. They either get washed away or are obliterated by subsequent ploughing and cultivation. For forming such bunds, an implement, called bund-former, was originally developed at Coimbatore. The implement consists of two concave blades, 90 to 150 cm long and 10 to 15 cm wide, joined across in a conversing manner. This implement is to be worked in a well-ploughed and harrowed field, just for collecting the soil and not for digging it. Its job is to collect the soil at the wider front and deposit it in the form of a ridge or a low bund behind. There are three sizes of bund-formers with blades 9 and 12 cm long. Recently a bigger bund-former, called "sara-making" machine, has also been designed. A wooden bund-former that would cost less than 10 rupees has been prepared at the Agricultural Workshop at Nilokheri (Haryana). Its blades are made of *babool* wood and its construction is so simple that the implement can be easily made by a village carpenter. If the wood is made available by the farmer, the implement would not cost more than 6 rupees.

The Forest Department in Madras has developed basin-listers for forming a series of a small basin or depressions in the land in order to collect rain-water. This is a useful operation in dry farming, because it conserves moisture. However, the implement, being rather costly, has not become popular.

Seed-sowing Devices

Among the indigenous implements, many seed-sowing devices are attached to implements, such as the *desi* plough. They are commonly called *nari* plough, *tifan*, *argada* and other multi-type implements that are used in the Deccan. Tube or tubes are fixed to these implements and on the top of these, a seeding-bowl, in which the seed is dropped, is fixed. The seeding-bowl has a number of holes in it. The construction of the holes is ingenious, with the result that the seed is dropped fairly evenly into the tubes and goes down to the tubes. In some areas, which are considered backward agriculturally, broadcasting or the sowing of seed at random on the cultivated field is also followed but the major portion of the area is nowadays sown in line and the seed is drilled. Sowing in lines ensures the proper distribution and placing of seed, resulting in the uniform spacing of the plants, with a concomitant reduction in seed-rate. It also helps the seeds to germinate regularly, ensuring a better stand of the crop and also facilitates inter-culture with bullock-operated hoes, thus reducing the cost of cultivation.

Quite a large number of improved seed-drills, both automatic and hand-operated, have been designed. Among the hand-operated ones, a copy of the wooden *tifan* has been made in Uttar Pradesh. It is made of mild steel, with an arrangement for changing the distance between the tines. The price of such a seed-drill is Rs 75 to 100. Among the automatic seed-drills mention may be made of the McCormic seed-drill, the Master seed-drill, the Gunti seed-drill and the Kirloskar seed-drill. The McCormic seed-drill uses an external-force feed-type of seed-metering device. The price of this seed-drill for a three-tined implement is Rs 1,200. The Master seed-drill has a series of small cups or spoons fixed on the discs. These discs revolve in a box filled with seed. The spoons gather the seeds and drop them into the tubes. The price of the Master seed-drill is Rs 950. The Gunti seed-drill also uses an external-force feed-type seed-metering device, but it has also a fertilizer-metering box fixed alongside the seed-box. The Kirloskar seed-drill which has recently been introduced has an ingenious device for seeding but it has not become popular. Most of the manufacturers of seed-drills are now making efforts to combine the simultaneous feeding of fertilizer with the dropping of the seed, thus making the implement a seed cum fertilizer drill. So far, however, none of these combined devices has proved very popular because of high cost, complicated designs greater draft, etc.

The seed-drills need a lot of skill in operation. It is necessary, particularly in respect of automatic seed-drills to calibrate them before sowing, so as to know the exact seed-rate that it will drop. The seed-drills are to be calibrated for each type of crop separately because of the differences in the shape, size and weight of the seeds of different crops. Similarly, in the case of the automatic seed-drills, proper care has to be exercised in choosing the soil-working part, such as the tine. If the soil is hard, strong straight tines with sharp edges in front should be selected. If the soil is soft, the disc shovels may be preferred. Some other important factors which a designer of a seed-drill and its user have to keep in mind can also be mentioned. These are the covering of the seeds, the pressing of the soil slightly over the seeds, the transport facilities for these drills, etc. Among the sowing devices, mention can also be made of transplanting-machines. Even though strenuous efforts are being made in different countries, such as Japan, Italy, the USA and China, so far no really useful machine for transplanting paddy has been prepared, except in Japan. Some of these new machines have been brought for trial at the Central Tractor Testing and Training Station, Budni, and have given quite good results as far as transplanting is concerned, but the machines require specially raised seedlings in trays and this is an elaborate and costly process. Therefore, transplanting has to be done by hand. A small tool for transplanting two or three seedlings of paddy at a time is being used in Burma and Malaysia. There is need to introduce this tool in India, because with little practice it can increase the efficiency of the labourers at least by 20 per cent.

Most of the interculture in India is done with a hand-tool, called *khurpi,* which consists of a sharp-edged triangular or sickle-shaped blade. Sometimes, weeding is done with a small spade, hoe or *mummuty.* Most of these tools have small handles and the user has to sit on his heel? or bend his back while weeding.

Weeding is one single operation which can increase crop yields without much additional investment, because efficient weeding will save moisture, crop nutrients and, to some extent, air and light required for the cultivated crop. To increase the efficiency of weeding, long-handled hoes, such as the Singh Hand-hoe, the Sharma Hand-hoe and the Dutch hoe have been introduced. The price of the hand-hoe is around Rs 6, including the handle and it is well within the means of even a poor farmer. It has been observed that four to five times more of weeding can be done with such long-handled hoes than with a *khurpi.* These hand-hoes are simple in construction and can be made by the village blacksmith. The use of these hand-hoes is, therefore, strongly recommended.

The wheel-hoes are a further improvement upon the hand-hoes. A wheel-hoe consists of a body, handle and a wheel. To its body, different attachments, such as shovels, a small plough, a small ridger, and other types of blades can be fixed. The price of the wheel-hoe is Rs 60 to 75, with a Weeding-attachment. Other attachments can be purchased at extra cost. Being provided with a wheel, it is easier to work and is very useful for inter cultivation in vegetable gardens and orchards. Where the soil is hard, one man can pull this hoe and another can direct it. Where the soil is soft, it requires only one man to work it.

For hoeing line-sown crops with bullock power a light *desi* plough is used in northern India. In the area between the Narmada river and the Krishna river, multi-tined bullock-operated hoes are used. They are known as *doura (Gunti), douri, dundia* and *dulari.* These are the miniature forms of *bakhars.* When the plants are small, multi-tined hoes fixed to a single wooden body are also used in Andhra Pradesh and northern Karnataka. The draft of these hoes is very light and sometimes two to four of them are pulled by one pair of bullocks. In the dry farming areas of Maharashtra, a *doura* with a split blade is used. The blades in this case work on each side of the row of plants. In the case of rain-fed crops, hoeing by using bullocks is done at the most three or four times, after which the plants grow tall and bushy and also the soil becomes hard. To cope with this condition, stronger and sharp hoes have been designed. They are commonly known as the Washim hoe, the Akola hoe, the Baroda hoe and the *triphali.* The designs of these hoes are more or less similar. They consist of three tines with a sharp triangular blade fixed to each. These hoes are very useful for the interculture of the cotton crop sown in rows.

Yet another implement called Indore Ridger, was introduced in Malwa some years back. This implement has a sharp beam, with a sloping ends and a hexagonal share attached to it with a strong iron bar. The ridger can be used up to a later stage of the crop without doing physical injury to the plants. It not only eradicates weeds but also produces a good mulch. Similar to the *doura,* two or three such ridgers can be attached to a pair of bullocks. It is surprising that such bullock-drawn inter-cultural implements have not yet been introduced into northern India despite their being used by the Indian farmers in the Deccan for centuries. There is a great deal of scope to introduce such implements into those areas where they are not being used at present.

For some special crops, such as paddy, special types of paddy-weeders have been designed and are being mass-manufactured in India. They are based on the Japanese design. They consist of two rollers with pointed fingers and are pushed in between the rows of the paddy plants when water is standing in the field. The idea of using these weeders is not to uproot the weeds and bring them to surface, as in the case of other weeders, but to bury them in the mud so as to suffocate and decompose them, and add organic matter to the soil.

With the advancement of scientific research, new methods to control weeds with selective weedicides, such as 2, 4-D, are being used. When this weedicide is sprayed in between the rows of a crop it kills the weeds only and not the crop. Sprayers are used for spraying these weedicides. At present, the weedicides are costly, but when their supply becomes abundant, the killing of the weeds with selective weedicides may become quite popular in India and it may not be necessary to use other implements to a great extent.

The other implements may have to be used for forming a fine mulch in between the rows, so as to conserve soil moisture. Some of the intercultural implements may have some attachments fixed to them to carry out some special jobs, such as sulphur-dusting in tobacco cultivation and the top-dressing of fertilizers. Already, simple sulphur dusters which can be worked in rows, and fertilizer attachments to the weeders have been designed.

Harvesting Implements

So far, the harvesting operation has not yielded to any mechanized tool. Almost universally, sickle is the only tool that is used for harvesting crops in India even though efforts have been made to introduce reapers, mowers, and even combined harvesters and threshers. These machines have, however, proved successful only on big estates or government farms, but because of their high cost, they have not become popular with the Indian farmers except when used on custom basis, as in Punjab and in Haryana during the last five years. The European scythe has also been tried, but except for cutting of lucerne or some such green fodder they have not proved very useful. One of the main reasons seems to be the shattering of the grain because of the impact of the harvesting implements and machines. In the case of sickles, there is less of shattering. The sickles are of two types, namely, plain or serrated or saw-shaped. The serrated sickles are used for cutting paddy and wheat.

Thrashing Implements

Threshing in, the case of paddy, if the area is small, is done mostly by beating the harvested plants with sticks. In the case of other crops, it is done by treading the crop with a team of animals. Stone rollers, about 90 cm in length and 30 to 60 cm in diameter, have also been used in some areas. However, no other thresher has become so popular as the "Olpad Thresher", which consists of 14 to 21 serrated discs fixed on a rectangular wooden frame in three lines. A seat is provided for the driver and also a guard. The implement is pulled by a pair of bullocks and worked on the harvested material which is spread-thickly on the threshing-floor. Threshing with this thresher is not only cheaper but is also quicker and more efficient than treading by bullocks and by using other methods.

The Japanese pedal-operated thresher, which was introduced into the rice-growing areas has not proved very popular because it does not give as much yield of grain as beating with sticks does. The reason for the Japanese using this type of thresher is that they are keen to get as long pieces of straw as possible in order to prepare ropes, mats or sacks out of the paddy straw. These articles are not used by the Indian farmers and, therefore, this thresher has not become popular in India. It can thresh about 300 kg of paddy in a working day of eight hours.

The threshing of wheat in some of the large private and government farms is done by trampling the harvested crop under a tractor or under a disc harrow attached to a tractor. It has been noticed that a 15-bhp tractor with a suitable harrow produces about 1,860 to 2,975 kg of clean undamaged grain per day of 10 hours at a working cost of Rs 35 to 40. The *bhusa* or the broken straw is also of good quality.

Recently, the Government of Punjab has introduced a power-operated wheat-thresher. It is worked with a tractor or an oil-engine of 20 to 25 bhp. This thresher is being simplified so that it can also be operated with an oil-engine of 10 to 15 bhp. The thresher is used by the owner alone if he himself is a big farmer, or can be given to other farmer, or can be given to other farmers on custom basis, or can be used co-operatively. In most of the villages in Punjab, Haryana, and western Uttar Pradesh, now at least one thresher exists, and in many cases there is more than one. Thousands are used on custom basis. The *bhusa* is a precious commodity and finds a ready market as a cattle feed. Therefore, a thresher, which cannot make *bhusa, w*ill not become popular. At the research stations in the country, the Japanese power-operated paddy-thresher is being given a trial. This machine can turn out 50 kg of clean paddy per hour. However, it has no *bhusa-making* arrangement. To make *bhusa,* the straw has again to be trampled under the feet of the bullocks. The drum-type power-operated thresher of 10 to 15 bhp has become very popular in northern India. This thresher is an example of a machine which the farmers in that region have readily adopted because of its great utility. Now virtually threshing by trampling straw by using, bullocks is limited to very small farmers. The introduction of this machine has saved much labour and has made threshing very easy and to complete it before the monsoon.

Different Winnowing-fans and Winnower Tools: To separate the chaff from the grain, the threshed material is put before a good natural breeze and the grain is allowed to drop through gravity and be separated from *the bhusa,* which being lighter, falls a little further away from the grain heap. In many areas, however, the natural breeze at the time of winnowing is not readily available and the farmers have to wait for hours and hours even for days to avail themselves of the natural breeze. Usually, in such areas, there is a danger of rain and hail-storm during this period. The farmers are, therefore, very anxious to take the season's produce to their homes as early as possible. Some crude methods of producing artificial breeze have been used by the farmers, but recently, winnowing-fans, similar to the ceiling fans, have been introduced. These fans may be operated by hand or with a pedal and the number of revolutions are increased by using a pair of gears or with sprocket and chain or a belt. In some cases, e.g. the Nilokheri winnowing-fan, such a drive is not necessary. The size of the fan itself is such that it can produce a good breeze. The object of these fans is to produce a breeze only and, therefore, no sieves are provided.

The winnowers are the machines with a series of three to four sieves to do the preliminary grading. A fan is enclosed in a box-like arrangement and is rotated fast. The sieves are also provided with a shaking arrangement. The steel winnowers cost Rs 1,000 to 1,500 whereas the wooden ones cost Rs 800 to 1,000.

Machines Tools Processing: Under this heading, the machines that process the farm produce during the post-harvest period have been described. These machines include sugarcane-crushers, fodder-cutters, groundnut-decorticators, paddy-hullers or paddy-shellers.

Crushers for Sugarcane: The growers of sugarcane cannot dispose of their entire harvest as cane. Part of it has to be converted into *gur* or jaggery. *Gar* is also a popular article of food in the villages. Formerly, the juice was extracted from the cane with crude two-roller wooden crushers. With the growing of improved varieties of cane, a three-roller cast-iron crusher operated either by bullock power or by engine power has been put into the market. It has greater crushing efficiency and, therefore, farmers are either buying such crushers or taking them on rent during the season.

Cutters for Fodder: Of all the improved agricultural implements, the fodder-cutter has become most popular, particularly in Punjab, Haryana, Delhi and Uttar Pradesh areas where green *jouar* or corn is required to be cut for feeding to the cattle. Two sharp steel blades are fixed to a circular flywheel. An automatic pushing arrangement is also provided. In some areas, where power is available, farmers are using the power chaff-cutters or fodder-cutters, or getting their chaff cut on rent basis. Chaff-cutters with good-quality blades are now manufactured in large numbers in India and are available at fairly reasonable prices. The power-driven chaff-cutters can be used also for cutting dry *jowar* stalks, which is not easy to cut with a hand operated chaff-cutter. For farmers, who own small holdings, the *Gopuri* type chaff-cutter designed at Sewagram near Wardha is recommended. It costs only Rs 40 to 50, whereas the cost of the hand-operated circular chaff-cutter is Rs 200 to 300.

Decorticators for Groundnut: When groundnut is to be sown, women labourers are employed for breaking the pods and taking out the kernels. In doing so, the testa of the kernel also gets separated to a considerable extent making it useless for sowing, because such damaged kernels will not germinate. Groundnut-decorticators, therefore, have been designed. The Maganwadi type of groundnut-decorticator is like a cradle with a screen placed below. The other rotary decorticators, either hand-operated or power-operated, are also available. The price of a hand-operated decorticator is Rs 200 to 300 arid the price of a power-operated one is Rs 600 to 1000

Hullers for Paddy: Paddy-hulling or paddy-shelling has been done in India with a *dhenkli* or wooden lever. The Khadi and Village Industries Commission has introduced a Rajasthan type of huller which is quite efficient. Recently, a manufacturer in India has put into market hand-operated hullers similar to Japanese types. They cost Rs 1,000 to 1,500, but are good for small holdings. They are operated by two persons with levered handles. The shelling or hulling is accomplished by the impact of the grain on a very fast-revolving rubber-disc.

A manufacturer has also put on the market a paddy-polishing machine. After the paddy is shelled or hulled, the grain is polished with this machine.

10

SCHEMES AND MANAGEMENT

Programm e planning is a procedure of working with the people in an effort to recognise unsatisfactory situations and to determine possible solutions or objectives and goals.

Kelsey & Hearne (1949):

"An extension programme is a statement of situation, objectives, problems and solutions."

Lawrence 1965:

"An extension programme is the sum total of all the activities and understanding of a country extension service. It includes

(1) The programme planning process

(2) Written programme statements

(3) Plan of work

(4) Programme execution

(5) Results

(6) Evaluation

Components of Schemes

The components of programme planning are complex. There are:

- People their needs
- Useful technology
- The educational process
- Analysing situation
- Making decisions about what should be done
- Determining action
- Projection the desired shape of things into the future etc. Principles of Programme Planning

For the development of a useful and effective programme, the following principles should be kept in mind, which is suggested by Kelsey and Hearne (1955).

(1) Extension programme planning is based on an analysis of the facts of the situation.

(2) Extension programme planning select problems based on the needs and interests of the local people.

(3) Extension programme planning determines priorities; defines objectives and solutions which offer satisfaction.

(4) Extension programme planning has performance with flexibility.

(5) Extension programme planning balance with emphasis.

(6) Extension programme planning has definite plan of work.

(7) Extension programme planning is an educational process.

(8) Extension programme planning is a continuous process.

(9) Extension programme planning is a coordinating process.

(10) Extension programme planning involves local people and their institutions.

(11) Extension programme planning provides for evaluation of results.

India Based National Plan

Basic Plans: These plans which were formulated before independence were

(a) *The Bombay Plan:* This plan was called " A plan of economic development of India." It was prepared by eight businessmen of Bombay in 1945. The main objective of this plan was to double up the per capita income within the period of 15 years. It was divided into 3 stages of 5 year each.

(b) *The People's Plan:* It was prepared by Shri M.N. Roy in 1944. Its guiding principle was " Production for use as distinct from production for exchange".

(c) *Visvesvaraya Plan:* This plan was prepared by Shri M. Visvesvaraya and accepted by the All-India manufacturers organisation in 1944. Its primary object was to raise the country's economic efficiency to a level at which the average citizen would be able to find employment and to earn a livelihood.

(d) *Gandhian Plan:* It was prepared by prof. S.N. Aggarwal, based on Gandhian ideas. Its objective was to raise the material as well as the cultural level of the masses to a basic standard of living within a period of 10 years.

Establishment of National Planning Committee: A national planning committee was appointed by the Congress working committee of India in 1938, with Shri Jawaharlal Nehru as its chairperson. The committee laid out certain guidelines so as to realise planned advancement from year to year.

Planning Commission: The planning commission is a multi-member body. The Prime Minister of India is the chairman of the planning commission. The Minister of Finance is the member in charge of finance in the commission in an ex-officio capacity. These arrangement ensure co-ordination between the commission and the central ministries.

Development Programmes for Villages

Various rural development programmes were started after in the year of 1960.

(a) *Intensive Agricultural District Programme (IADP):* The intensive agricultural district programme, popularly known as the package programme, was started on a pilot basis in 1961 in 7 selected districts. The programme goals at combining technical know how credit and production supplies for stepping up agricultural production.

- The operation of IADP has contributed to a significant improvement in the use of critical inputs like improved seeds, fertilisers and plant protection measures.

(b) *High Yielding Varieties Programme (HYVP):* The high yielding varieties programme was started in 1966-67. The cultivation of HYVP has resulted increase in food grains production. Wheat production has been more than doubled. Rice production has also increased through not as much as in the case of wheat.

- A major development in rice cultivation was the launching of the Minikit programme from 1971-72.

(c) *Drought Prone Areas Programme (DPAP):* The drought prone areas programme formerly known as rural works programme was initiated in 1070-71 as a non central sector scheme. The main aimed at mitigating the severity of scarcity conditions by executing rural works to generate employment.

(d) *Integrated Rural Development Programme (IRDP):* Integrated Rural Development Programme is a non-plan programme which means IRDP comes under five year plan.

- The main aims of IRDP is the all round development of family.
- IRDP was launched on 2nd Oct. 1980 by integrating following programmes

(i) SFDA -Small Farmers Development Agency

(ii) DPAP—Drought Prone Area Programme

(iii) Operation Flood II

(iv) NREP—National Rural Employment Programme

(v) MNP—Minimum Need Programme

District Rural Development Agency

The DRDA (District Rural Development Agency) is headed by the collector/ deputy commissioner or CEO zila parishad in states.

- DRDA helpful for Implementing National Rural Employment Programme (NREP). National Agricultural Research Project

NARP was launched in January, 1979 and is founded by the ICAR and through IBRD. The soft loan connected of the world bank.

- The NARP is the first major effort of the Indian Council of Agricultural Research (ICAR) to promote and strengthen the regional research capability of the State Agricultural Universities (SAUs)
- The project is designed to fill the gaps in research.

NAEP: NAEP (National Agricultural Extension Project) is being launched by the Government of India. So that on effective and efficient transfer of technology is assured. The links establised between NARP, IRD, and T and V system together produced the desired results.

- NAEP is being launched with the help of IBRD- World Bank connected to overcome the various constrains identified earlier and will rectify many of them.

T and V System: Training and Visit System is a new extension service approach. The concept of T and V system was evolved in 1973 by Israeli extension expert Daniel Benor. Therefore T and V system is also called Baster and Benor scheme and firstly evolved in Turkey (in Seyhan irrigation project)

- T and V system has two phages/stages: (i) Training, (ii) Visit
  (i) Training: Training is imparted for transfer of technical know how from subject matter specialist to extension worker.
  (ii) Visit: Extension worker has to visit to the farmers for transfer of know how obtained at the training.
- T and V programme is Govt. sponsored rural development programme.
- T and V system is introduced in India in 1974 in Rajasthan canal area and chambal command area in M.P. to improve the effectiveness of agricultural extension.
- Training and visit programme also called Intensive Agricultural Extension.

Launching of Lab to Land Programme

Lab to land programme was launched in India on the occasion of ICAR golden jubilee celebration year in July 1979 for only one year. But later on this programme was extended further year after year.

- That aims of lab to land programme is the transfer of proved and viable low cost agricultural technologies to small and marginal farmers and landless laboures to improve their overall socio-economic condition.
- That's why lab to land programme was launched under lab to land or land to lab programme, the problems related to the agriculture were brought to the laboratory where the particular solution was to be found out or searched out and later on were communicated to the farmer.

Krishi Vigyan Kendra (KVK): Education Committee (1964-66) recommended to establish agricultural polytechnique.

At last ICAR mooted the idea to establish Krishi Vigyan Kendra (KVK) and Mohan Singh Mehta Committee was appointed in 1974 by ICAR to formulate KVK scheme.

- First KVK was started at Pondichery in 1974 with the aim of transfer of technical literacy to increase agricultural production.
- The main purpose of KVK is to impart training at the Kendra or out of the centre.
- There were 314 KVK in number by the year March 2001 and plan is to establish one KVK in every district of India.

The credit for the success of KVK goes to Dr. Chandrika Prasad.

Establishment of Community Development Programme: In community Development programme, chief role was played by grow more food campaign. On 2nd Oct. 1952, 55 Community Development Project were started in different parts of the country with Indo USA agreement for three years. This project covered nearly 25,260 villages and a population of 6.4 millions.

- The basic idea and concept of community development was borrowed from USA.

Beginning of Panchayati Raj System

Panchayati Raj was first started at Nagaur Rajasthan and Andhra Pradesh in 1959. As a whole state Rajasthan was covered first by the Panchayati Raj system.

- In the Panchayati Raj system, village Panchayat is the lowest unit. The General body of Panchayat is Gram Sabha.

TRYSEM Scheme: Training of Rural Youth for self employment. TRYSEM was launched in 1979 with the basic principle of knowledge by doing or learning by doing to provide technical skill to the rural youth from families below the poverty line (BPL)

- The target group comprises rural youth between the age of 18 and 35 years old.

(Ins, PCS, 2000, 2001)

Establishment of 4H-Club: 4H-Club was established in 1900 in America to provide training to youth for agricultural and home development, it was first govt. step.

- 4H-Club means Heart, Head, Hand and Health was explained by A.B. Graham.
- On the basis of 4H-Club of USA, young farmer's association of India was started by P.S. Deshmukh in 1956.

Important Abbreviations

I.R.D.P.	—	Integrated Rural Development Programme
T.T.C.	—	Trainer's Training Centre
A.U.	—	Agricultural Universities
C.B.	—	Cooperative Bank
F.T.C.	—	Farmers Training Centre
W.V.D.P.	—	Whole Village Development Programme
K.V. K.	—	Krishi Vigyan Kendra

(ICAR, RAS 1999; JRF 2001 ; ARS 2001; U.P.PC.S. 2000)

T.A.D.	—	Tribal Area Development
H.A.D.	—	Hill Area Development Programme
M.F.A.L.	—	Marginal Farmers and Agricultural Labour Programme
I.D.A.D.	—	Integrated Dryland Agricultural Development
D.PA.P.	—	Drought Prone Area Programme
S.E.D.A.	—	Small Farmers Development Agency

(PCT 2003)

C.A.D.	—	Command Area Development
C.D.D.	—	Cattle and Dairy Development

S.D.	—	Sheep Development
P.D.	—	Poultry Development
I.A.D.P.	—	Intensive Agricultural Development Programme *(U.P. PCS 2002, IAS)*
I.A.A.P.	—	Intensive Agricultural Area Programme
H.Y.V.P.	—	High Yielding Varieties Programme
I.O.D.P.	—	Intensive Oil Seed Development Programme
I.C.D.P.	—	Intensive Cotton Development Programme
N.C.A.	—	National Commission on Agriculture
ICRISAT	—	International Crop Research Institute for Semi-Arid and Tropics
LDB	—	Land Development Bank
ARDC	—	Agriculture Refinance and Development Corporation
JRY	—	Jawahar Rozgar Yojna
FCI	—	Food Corporation of India
NAFED	—	National Agriculture Cooperative Marketing Federation
IFFCO	—	Indian Farmers Fertilizer Cooperative Ltd.
AGMARK	—	Agricultural Marketing
RRB	—	Regional Rural Banks
FSS	—	Farmers Service Society
SBI	—	State Bank of India
IFCI	—	Industrial Finance Corporation of India
UTI	—	Unit Trust of India
IDBI	—	Industrial Development Bank of India
LBS	—	Lead Bank Scheme
WTO	—	World Trade Organisation
IMF	—	International Monetory Fund

NABARD — National Bank for Agriculture and Rural Development

NCDP — National Cooperative Development Corporation

AMA — Agricultural Marketing Advisor

APC — Agriculture Price Commission

NSC — National Seed Corporation

GNP — Gross National Product

TRYSEM — Training of Rural Youth for Self Employment

FAO — Food and Agricultural Organisation

CDP — Community Development Programme

Process of Communication

Individual

(1) Farm and Home Visit or personal contact

(2) Letters (Personal)

(3) Telephone call

(4) Office call

(5) Result demonstration

Group Contact Method: Group means a group of 2-30 persons:

(1) Symposium (2-5 speakers)

(2) Pannel (2-8 speakers) (3) Discussion

(4) Lecture (5) Tour and field days

(6) Method demonstration and result demonstration

(7) Meeting (8) Role playing

(9) Puppet show (10) Slide

(11) School (12) Conference

(13) Seminar (14) Training camp

(15) Public address equipment

(16) Flash card

Mass or Community Contact Method

(1) Bulletins	(2) Newspapers
(3) Journal	(4) Magazine
(5) Exhibition	(6) Circular letter
(7) Fairs	(8) Posters
(9) Cinema (Film)	(10) Movie
(11) Drama and songs	(12) Radio
(13) Leaflet/ folder	(14) Television

Glossary

- Aim: Aim is a broad objective
- Audio-visual Aid: Audio-visual aid is an instructional device that can be heard and seen.
- Audience: Audience is the intended receiver of messages.
- Association: It is a group of persons organising themselves for the purpose of fulfilling certain interests.
- Adopter: Adopter is a person who has continued full use of innovation.
- Andragogy: It means art and science of helping adults to learn.
- Agricultural Extension: Agricultural extension is the assistance to farmers to help them to identify and an their production problems and to become aware of the opportunities for improvement.
- Audio-aids: Audio-aids are those which facilitate understanding of a problem or topic through hearing.
- Agricultural Information Centre: Agricultural information centres are set up at district head quarter. It collects information suitable to the farmers.
- Black Boards: Black board is one of the most widely used visual aids which can be easily prepared everywhere with the use of locally available material and resources.

- Calender of Work: It is a plan of work arranged chronologically according to the time, when each step of work is to be done.
- Circular Letter: Circular letters are those which are written by persons who are at the high helm of affairs such as C.M., important public leaders.
- Communication: Communication is a process by which two or more people exchange their ideas, facts, feelings or impressions in ways that each gains a common understanding of the meaning and use of messages.
- Channel: Channel is the physical bridge between the sender and the receiver of a message.
- Cooperatives: These are the institutions organised by groups of individuals for mutual help, thrift and benefit.
- Community: Community is a sub group having many of the characteristics of society but on a smaller scale and with less intensive and co-ordinated common interest.
- Continuous Education: Continuous education can be given through both formal and informal types of education as and when need arises to educate for new technological development and for socio-cultural development.
- Contract Farmer: A farmer who act as intermediary between follower farmers and the extension education agency.
- Costoms: Costoms mean uniformly accepted ways of acting about some social aspects of life.
- Caste: When class is somewhat hereditary it is called as caste.
- Culture: Culture is the sum total of the ways in which human beings live and is transmitted from generation.
- Charge Agent: A professional person who attempts to influence adoption decisions in a direction he feels desirable.

- Demonstration Centres: These provide effective communication about improved methods of farm practices to the farmers by practical displays.
- Decision Making: Decision making refers to the process in which programmes are prepared by various departments in consultation with various persons.
- District Agricultural Officer: District agricultural officer is a person who makes research findings known to farmers for their use.
- Dramatics: Dramatics are a great source of entertainment and education and have been known to have a considerable effect on the society from times immemorial.
- Distance Education: It is any one of the various forms of the study which are not under the continuous immediate supervision of tutors present to their students in lecture rooms.
- Demonstration: Demonstration is conducted by farmer under direct supervision of extension workers to prove the advantages of recommended practices.
- Demonstration Plot: It is a small area set aside for result demonstration of extension staff.
- Extension: Extension is a method of non-formal education aiming at inducing behavioural changes in its clients for their increased knowledge, skills and income.
- Education: Education is the process of bringing desirable changes in human behaviour.
- Extension Education: Extension education is an applied science consisting of content derived from research, accumulated field experience and relevant principles drawn from behavioural sciences, synthesised with useful technology in a body of philosophy, principles, contents and methods focussed on the problem out of school education for adults.

- Extension Worker: Men and women employed by an extension agency or technical and professional staff employed at all level.
- Evaluation: Evaluation is the study of the degree of success, we get in terms of our original objectives of the programme.
- Extension Agent: The extension worker at the village level in direct day to day contact with local farmers.
- Folk Ways: Folk ways are customary ways of behaving in which society exerts some what, they are not rigid.
- Functional Literacy: Functional literacy is the ability to read, write and calculate, so that a person may engage in all.
- Farming Need: Farming needs are the gap between farming practices adopted by the farmers at present and the farming practices recommended by the research worker at present.
- Field Trip: Field trip is a group of interested farmers and extension education workers who go on tour to see and obtain first hand knowledge of improved practices.
- Flow Charts: Flow chart shows the functional relationship e.g., organisational set up in a block, district or state.
- Film Strips: These are the series of still picture in one roll.
- Folk Songs: Folk songs are a powerful means of propagating the idea and experience of a wise and the learned.
- Flannel Graph: It is a piece of flannel or sand paper stick on flannel.
- Forum: Two or more speakers who present talks on the same subject. The primary characteristics of the forum

is that the subject is controversial and speakers present the opposite sides of questions.

- Family: Family is a system of organised relationship involving workable and dependable ways of meeting basic social needs. It is a most multifunctional of all institutions in society.
- Flash Cards: These are small cards of thick paper with pictures, diagrams or drawings on a particular subject or topic.
- Films: Films are a series of motion pictures projected in sequences with a synchronised sound system.
- Group Meeting: Group meeting are a device to save time by several individuals simultaneously.
- Goal: Goal is the distance in any given direction proposed to be covered in a given period of time.
- Graph: Graph is a diagrammatic presentation of a quantitative data in a clear, interesting and easily understandable form and is very effective in showing trends, developments and relationships.
- Individual Meeting: These are between two or three individuals arranged at home or at the field.
- Integrated Rural Development: IRD is defined as development which can help to increase the purchasing power of the rural poor through the generation of greater opportunities of gainful employment.
- Innovation: Innovation is an idea perceived as new by the individual.
- Innovativeness: It is the degree to which an individual is relatively earlier in adopting new ideas than the other members of his social system.
- Interest: Interest refers to the attention with a sense of concern focussed on some objects.
- Illiteracy: Illiteracy means inability to read and write.

- Illiterate: It means person who has not acquired basic skills in reading, writing or arithmetic.
- Institution: Institution is a system of relationships or a pattern for carrying out an idea or a desire which is regarded as necessary for the welfare of the group.
- Learning: Learning is the process by which an individual through his own activities becomes changed in his behaviour.
- Lecture Note: Lecture note is merely a store of information.
- Lesson Plan: Lesson plan is an outline of activities that the teacher will follow in order to create effective learning situation for the learners. It is a plan for teaching.
- Leadership: Leadership is a process of influencing the behaviour of the individual in a real situation.
- Leaflets: Leaflet is a single sheet of printed matter folded or unfolded giving informations on a particular topic.
- Leader: An individual is a leader in any social situation in which his ideas and actions influence the thoughts and behaviour of others.
- Learning Experience: It is a mental/physical reaction, one makes through seeing, hearing or doing the things to be learnt through which one gains meaning and understandings in solving new problems.
- Method Demonstration: Method demonstration are those in which a new method, or technique is demonstrated, taught or given practice of.
- Motivation: Motivation is the process of creating a strong desire for appealing to the fundamental wants of the people so that they are influenced to change the behaviour.
- Models: It is the replicase of real objects or sealed representation of things.

- Mass Meeting: These are the meetings of a general mass people called to hear a leader or endorse the decisions of an executive committee formed by the general body of an organization.
- Monograph: Monographs are a technical publication, written on a particular subject.
- Message: It is the information, a communicator wishes his audience to receive, understand, accept and act upon.
- Newspapers: Newspapers are an effective medium for disseminating information about local, district, provincial, country and world affairs.
- Need: Need is a difference between two situations that "what is" and what ought to be."
- Norms: Norms refers to the rules that govern action directed towards achieving values.
- Objective: Objective is the direction of movement.
- Overtime: Over time refers to the time required to adopt from its origin.
- Opinion Leaders: Those individuals from whom others seek the formation and advise.
- Organisation: Organisation is a set of two or more parts which are so related that they function as a whole.
- Plan of Work: It is an outline of activities so arranged as to enable effective execution of the entire programme.
- Project: Project refers to an outline of procedure and portions to some phase of extension work.
- Personal Example: It is one of the most effective ways of training and communicating with the people. It is generally what we do and how we do it that makes a lasting impression on trainer or people.

- Programme: Programme is a systematic statement of situation, objectives, problems and solutions. It is relatively permanent but requires constant revision.
- Pamphlets: Pamphlets are very valuable and effective written material or literature for use in extension.
- Programme Planning: It is a process of working people in an effort to recognise the problems and determine possible solutions.
- Planning: Planning is a process of preparing systematic statement of the line of action for achieving certain overall and specific objectives in relation to needs and resources.
- Panchayati Raj: Panchayati raj refers to a members or an assembly of elders who settles disputes, according to custom in a village or within a caste.
- Poster: Poster is an important visual aid helping to arouse people.
- Psychology: It refers to the use of animals in behavioural research and collect information and in sight relevant to their area of concern.
- Panel: Panel is also termed as round table discussion. It is usually composed of 4-6 persons. The members of the panel carry on a discussion of the topic among themselves and the audience listens in on the discussion.
- Poverty: Poverty is the social phenomena in which a section of the society is unstable to fulfil even its basic necessities of life.
- Personality: Personality is a pattern body of inherited traits.
- Prestige: Prestige is the evaluation of the status.
- Result Demonstration: These are those in which two practices or techniques are compared for results.
- Rural Development: Rural development is a process of change form the traditional way of living to progressive way of living.

- Reference Group: Reference group is a group of persons whom an individual consult before taking an important decision.
- Role: Role refers to the function of the status.
- Rural Sociology: Rural Sociology is the systematic study of the people living in rural areas and follow patterns of occupation and life somewhat different than those living in urban areas.
- Staff Meeting: These meetings are among formal groups staff meeting occupy a unique position. They are an essential part of a successful extension programme.
- Subject Matter Specialist: Professional staff who provide technical back up to extensionist.
- Social Days: Social days are days on which the extension and the administrative personnel of the block and the district get together with the villagers in some village.
- Social Problems: It is a condition in society that is considered harmful or undesirable by society as a whole based on existing social values.
- Social Centre: The adult education classes will be organised through different adult education centres on the basis of needs and requirements of adult men and women.
- Social Class: Social class is an abstract category of persons arranged in levels according to the social status they posses.
- Social Group: Social group is a unit of two or more persons in reciprocal interaction or communication with each other.
- Social Value: These are defined as the attitudes held by individuals, group or society as a whole as to whether material or non-material objects are good, bad, desirable or undesirable.
- Social Stratification: Social stratification is the pattern of super imposed categories of differential privileges.

- Social Sanction: Social sanction is a reward or punishment applied to persons in order to encourage or discourage certain types of behaviour.
- Social Welfare: It refers to a large number of activities connected with the uplift and welfare of the weak and exploited sections of the society.
- Social System: A population of individuals who are functionally differentiated and engaged in collective problems and solving the problems.
- Status: It is a position or place in a social set-up.
- Social Control: Social control is the process of constraining persons to conform to various norms.
- Sociology: Sociology is the scientific study of men's behaviour in relation to other groups and individuals with whom he interacts.
- Specimen: It may be a part of an object or one of group or class to represent the whole group.
- Supervision: It is the process of placing into action standards of performance developing and utilizing human resources to achieve organisation results.
- Society: Society refers to the group within which men share.
- Television: Television is one of the most powerful audio visual aids. It combines in itself both visual and auditory aids and is very effective in conveying actual life events in action.
- Traditions: The uniformly accepted ways of thinking.
- Time Charts: They present data in ordinary sequence e.g., year to year increase in the income of the people, profit and loss account of farms etc.
- Technical Workers: These workers are trained in special fields. Their training includes some basic training in specialized fields followed by job training.

- Teaching: Teaching is the process of arranging situations that stimulate and guide learning activities in order to bring desired changes in human behaviour.
- Visual Aid: These are aids which facilitate understanding of a problem or topic through seeing.
- Village Level Worker: The multipurpose agent of India's community development programme at village level.
- Wall-newspapers: These are big sheets of paper with current news, experiences, recommendations etc., printed on them.
- Walking Tours: Walking tours are an important means of communicating with people.

11

New Trends and Techniques

Although agricultural engineering is a fairly new subject, its effect is being felt in India increasingly. In recent years it has contributed considerably to agricultural production in the country both directly and indirectly. The number of modern machines used in agriculture is a fairly good index of the progress made during the four Five-Year Plans. Roughly, more than 1,00,000 tractors, crawler-tractors and power-tillers are in use in India.

Similarly, the number of combines, self-propelled, trailed and mounted ones, has gone up to more than 1,500. The number of power-threshers may be in the vicinity of 5,00,000 and the number of both diesel and electric pumps must have crossed the million mark. And yet all this progress mostly made during the last 25 years is insignificant as compared with the use of modern agricultural machines in the developed countries, such as the USSR, the USA, Japan, the UK, Canada and Australia, and we have yet to cross a number of hurdles, but the progress made so far is quite encouraging and forms a basis for a leap forward. This development has, however, made two things clear:

(1) that the Indian farmers, though belonging to a less developed country, are quick in adopting new methods, techniques and machines, provided they are convinced of their utility, and

(2) that the Indian industries can manufacture as good modern machines as those manufactured elsewhere. The confidence created by these developments augurs well for taking bold steps in the future. And this in itself is no mean achievement.

Agricultural engineering, as defined by international institutions, comprises four main branches, namely:

(i) farm implements and power,
(ii) rural structures,
(iii) soil conservation, drainage and irrigation, and
(iv) rural electricity.

Because of the peculiar local conditions, the lack of development of electricity on a large scale, and the immediate utility of implements, agricultural engineering in India has come to mean more of agricultural implements and machinery. But, in fact, agricultural engineering is a very vast subject. And may be defined as the application of the knowledge, techniques and disciplines of various fields of engineering to the solution of problems arising in the fields of agriculture and rural living, with the object of reducing labour, improving agricultural productivity per worker, and raising the standard of living of the farmers and increasing the overall earnings per worker.

During the Second and the Third Five-Year Plans, agricultural engineering divisions were added to the departments of agriculture in the states and at the Centre. They undertook the work of hiring or selling tractors, oilengines, implements, pumps, etc. Several training institutions and workshops were started to train artisans and mechanics, and in every state a research cum testing and training centre for agricultural implements was established. The cumulative effect of all these steps has been to popularize mechanization and to prepare a sound base for it.

Then came an era of more intensive planning and re-organization on a wider scale. With the establishment of agricultural universities, several agricultural engineering colleges were established, offering graduate and post-graduate courses. In most of the states, where such colleges were established, they took over research, education and extension from the state departments of agriculture. The most important step in the reorganization has been the establishment of agro-industries corporations, one in each state, with the financial assistance of the Central Ministry of Agriculture. The main objectives of establishing these corporations have been to take over the supply and service functions of the departments of agriculture and to expand them. These agro-industries corporations have each a paid-up capital of 2 to 5 crores of rupees. They have taken over the existing government workshops also. The progress made by these agro-industries corporations is given in this chapter.

The other most important development has been in the field of industry which has established so far half a dozen factories for manufacturing tractors, 3 or 4 for manufacturing power-tillers and several other factories for manufacturing oil-engines and electrically-driven pumping-sets. Some of these industries are being substantially assisted by the Industrial Finance Corporation of India. The Government has also established a factory for manufacturing crawler tractors and another one for manufacturing wheel-type tractors. These three main developments helped to quicken the pace of mechanization of Indian agriculture.

The following tabulated statement clearly shows the progress made during the last 25 years:

Sl. No	*Name of the Implements*	*1951*	*1956*	*1961*
1.	Mouldboard-type iron ploughs	9,30,000	13,66,000	23,00,000
2.	Bullock-driven sugarcane crushers	5,04,000	5,44,000	5,90,000

Contd...

Sl. No	*Name of the Implements*	*1951*	*1956*	*1961*
3.	Power-driven sugarcane crushers	21,000	23,000	32,210
4.	Centrifugal pumps with oil-engines	82,000	1,22,000	2,29,000
5.	Electric pumps	25,000	55,000	1,60,000
6.	Tractors	8,500	21,000	34,000

It is not merely the knowledge of agricultural engineering, including implements and machinery, which is necessary for increasing agricultural production, but also the knowledge of the sources of their availability, prices and upkeep. The following is a list of sources through which the information on the availability and prices of the machinery can be obtained.

(1) *At the Village Level:* The Village-Level Worker or *gram sewak.*

(2) *At the Block Level:* The Agricultural Extension Officer or the District Planning Officer.

(3) *At the District Level:* The District Agricultural Officer or the District Planning Officer.

(4) *At the State Level:* The State Agricultural Engineering Section and the Managing Director of the State Agro-Industries Corporation.

(5) *At the Central Government Level:* The Director of Agricultural Implements and Joint Commissioner (Machinery), Ministry of Agriculture and Irrigation, Krishi Bhavan, New Delhi. The private industry as well as the government, through their agro-industries corporations, are in the field now for the supply of agricultural machinery to the farmers. To prevent the charging of exorbitant prices for tractors, the government has fixed the maximum prices of tractors

with the assistance of the Bureau of Cost Accounting in the Ministry of Finance. This organisation has helped the farmers very much.

Future Prospect: Agricultural aviation is now gaining ground in India. At present, the aeroplanes and helicopters are being used mostly for spraying pesticides, but in future, they may be used for emergency sowing, and spraying liquid fertilizers on the growing crops. Agricultural electronics has also begun and the remote-control tractors are being designed and tested in some countries. Japan has successfully designed machines for transplanting paddy seedlings and their trials are going on at the Tractor-Training and Testing Centre at Budni. A very ambitious programme for joining the Ganga with the Kauveri, via the Mahanadi, the Narmada, the Godavari and the Krishna, is under the consideration of the Central Government and some preliminary work is being carried on. With the successful explosion of an atomic device, India may be in a position to generate more electricity in future. The availability of plentiful irrigation water as a result of the Ganga-Kauveri Project, and the availability of more electricity from atomic energy will make the progress of farm mechanization in India speedier. Thus the future seems to be bright.

Machanical Aids

The beginning of mechanization of Indian agriculture was made by the use of improved hand-tools and improved bullock-drawn implements. Particular stress was laid on these items during the first three Five-Year Plans and slowly more sophisticated implements were introduced during the Fourth and Fifth Plans. These included power driven threshers, tractors and the combine harvesters. Since simpler implements were taken up first and since in many areas they are still being used, their description is taken up here first. Quite a variety of implements and machinery is required for carrying out various agricultural operations, starting from the opening of the land to harvesting, threshing, winnowing and storage. These operations include ploughing, harrowing, levelling, sowing, intercultivation, the application of fertilizers and manures, harvesting, threshing and winnowing.

India is a sub-continent and a variety of indigenous or *desi* implements are in use in different parts of the country for the last so many centuries. Before efforts were made to introduce foreign implements, it was necessary that a survey of indigenous agricultural implements used in India should be undertaken. Therefore, under the Indian Council of Agricultural Research, a country-wide survey was carried out and its report was published. This survey has brought a number of important facts to light. The number of implements used by the Indian farmers is really very large. Some of these implements, even if they look crude, are really useful and the consensus of the agricultural engineers in India has been that it would be easier at least as a first step to improve upon the indigenous implements in order to increase their efficiency and thus give to the Indian farmers something which is within their means.

To achieve this objective effectively, the Indian Council of Agricultural Research has now established one research-cum-testing and training centre in each state. Some of them have now been converted into regional centres or have been taken over by the state governments. The object is to improve the indigenous agricultural implements, to design new implements and to test them in the field up to the prototype level. After the implements have proved to be useful, the prototypes can be given to the manufacturers both in the public as well as in the private sectors and the implements can then be sold through them to the farmers.

Research Principles: In India, the holdings are generally small, the average farmer is poor, and the draft cattle are also small and ill-fed. Taking these factors into consideration, certain principles, on which research is to be carried out, have been laid down. These are given below:

(1) That the implements and machines for the Indian farmers should be simple in construction so that they could be operated by the illiterate farmers, and should

either be manufactured or could be repaired by the village artisans or mechanics.

(2) The prices of the implements should be within the reach of the ordinary farmers, though they need not necessarily be cheap at the cost of efficiency, but they should also not be too costly. In the case of power-driven equipment, the difficulty about the cost can be overcome by giving the machinery on hire or hire-purchase system or on custom basis or can be utilized on co-operative basis.

(3) The implements should be light, so that they can be transported from the village to the field and that the draft of the implement is suitable for the types of animals that are used in particular areas.

(4) That the implements should be scientifically tested on a fairly uniform basis throughout India, so that the results are known and can be compared. There should be as little guess work as possible.

(5) In designing the implements, as far as possible, the local available materials are to be used. In the case of soil-working and plant-working parts, durable materials, such as high-carbon steel, should be used.

To fulfil these requirements, the Indian Council of Agricultural Research has also finalized procedures for the testing and designing of implements and the proformas in which the test data are to be collected. In recent years, the attention of the Indian farmers has been drawn to the power-operated machines and equipments, such as tractors, oil-engines, and pumps. Quite a large number of the farmers are using these implements and firms have also been established in India to manufacture them. This trend towards mechanized farming is bound to increase.

Recently, however, new developments have taken place to bring about mechanization of agriculture. Owing to the shortage of diesel-oil and its increased price, the mechanical operation will have to be restricted to deep ploughing, land-levelling, clearing land, and other operations which cannot be carried out with bullock-power. Now greater attention has to be paid to the bullock drawn implements and manually operated implements. For instance, for short distances, bullock-carts have been found more economical than trucks; hence the necessity for improving bullock-carts and also bullock-drawn, but improved implements, on priority basis. The description of some improved implements is given in the following pages:

Ploughs: A plough is the primary implement used all over India. The *desi* ploughs have almost a common design and they differ mostly in size, depending upon the type of the soil and the size of the bullocks or buffaloes used in the particular area. The heaviest ploughs are found in the black-cotton-soil areas of India. The *desi* plough is almost like a western-style cultivator, both in design and action. The body and the wedge may be of one piece or of two pieces. The iron point or share is made up of a rod or bar of mild steel, and penetrates into the soil and protects the wooden wedge against wear and tear. The beam varies from 2.5 to 3.5 m in length and is made of wood. The handle is fixed to the beam at the back of the body. All indigenous ploughs are almost similar in design, but vary considerably in size and weight to suit the prevailing soil types, ploughing conditions and the weights and heights of the draft animals in the locality. The working draft varies from 27 to 148 kg. The following table gives an idea of the size and the draft of the *desi* ploughs used in different states.

The *desi* plough makes a V-shaped furrow and, unlike a mouldboard plough does not turn over the soil. It is mostly made of hard wood, e.g. *babool* or *kikar* (*Acacia nilotica*). The depth of ploughing is usually from 10 to 2.0 cm and the width from 13 to 18 cm. There are minor variations in design, weight, the length of the beam, etc. The duck-foot type of plough, as used in Orissa is different in design. In has a flat bottom of about 38x30 cm and is said to be useful under soil conditions in Orissa. The plough used in Maharashtra is very heavy requiring sometimes 2 to 4 pairs of bullocks. It is useful for ploughing the black cotton soils of the Deccan. Most of the *desi* ploughs are light in weight and can be carried either on the yoke of the bullocks, or on the shoulders by the farmers.

The depth of the plough can be adjusted with a wedge provided in the body of the plough or by adjusting the yoke either forward or backward on the beam. If the yoke is fixed forward, the plough goes deeper and if the yoke is taken backward, the plough goes shallower. In Andhra Pradesh, a simple triangular plate is fixed in front of the body of the plough so that ridges and furrows can be made. This is an ingenious device to convert the *desi* plough into a ridger. Since the *desi* plough acts like a cultivator, repeated ploughings and cross ploughings are necessary to secure a good tilth. The *desi* ploughs are also used for making shallow furrows for depositing seeds, for inter cultivation between the rows of crops and to harvest crops such as potatoes, groundnuts and sweet-potatoes. It can thus be seen that the *desi* plough is a multipurpose implement difficult to be replaced, unless more efficient multipurpose improved implements are designed.

Weight, Draft and the Furrow Size of Different Desi Ploughs

Ploughs	*Weight (kg)*	*Length of bean (m)*	*Average draught (kg)*	*Horse power expended*	*Average size of furrow (cm)*	
					Width	*Depth*
Bareilly	10.88	2.66	28.57	0.40	17.78	10.16
Pusa (Bihar)	15.42	2.5	27.66	0.43	26.4	17.78
Angul (Orissa)	10.88	2.15	47.17	0.73	27.94	15.42
Sabour (Bengal)	19.95	2.87	100.24	1.35	30.4	20.32
Patna	15.87	2.71	54.43	0.81	27.94	20.32
Kerala	23.58	3.32	134.26	1.87	30.4	20.32
Dohad (Mumbai)	13.15	2.84	158.75	2.67	22.86	22.86
Coimbatore	37	3.53	328.0	1.71	27.94	22.00
Indore	57	3.15	403.5	1.71	22.00	26.00

Bladed Harrows: These are called *bakhars* or *guntakas. Bakhar* is a very common implement used in the Deccan. It is one of the best implements for primary tillage. The farmers use it for preparing the land and for sowing the seed. It consists of a steel blade of about 45 to 90 cm in length and 8 cm in width. It can cultivate the land to a depth of 9 to 10 cm. Even the dryland tillage is done with this implement by adding a wedge to it. If the soil is very dry, a man stands on the implement to give it the weight. It can cover 0.8 to 1.6 ha in a day because of its greater width. The Lanka *bakhar* used in the Vidarbha area has a blade even up to 1.5 m in length and is used after a few monsoon showers to kill young germinated weeds. The implement is worked cross-wise in the fields to achieve a better tilth. This is a very useful implement and costs only Rs 30 to 45. Somehow, this implement has not yet been introduced into areas above the Narmada river, except in some pockets near Jhansi and Gwalior. It needs to be examined whether such an implement can be introduced into Bihar, Uttar Pradesh and Punjab. Almost every farmer in the Deccan possesses this implement. It is interesting to mention here that in some parts of the USA, where moisture conservation is an important problem, the use of "sweeps" and "chisel cultivators" which are similar to the Indian *bakhar* or blade harrow in working, is being advocated. The Egyptian and West Asian ploughs are also very much like the Indian bladed harrow. Recently, improvements have been made in the design of the *bakhars,* making it possible to change the blade as desired. They are mostly made of steel.

Mouldboard an Improvement in Plough: A mouldboard plough consists of a share and a mouldboard of steel. This is probably one of the first implements to be introduced into India from foreign countries. It differs in action from the *desi* plough in ploughing a square furrow and in inverting the soil either completely or partially, depending on the curvature given to the mouldboard. For a proper functioning of the mouldboard ploughs, the provision of an adequate vertical and horizontal suction to the share up is necessary. In the case of long-beam ploughs, the beam and the handle are made of wood, whereas in the case of a short-beam plough, the complete implement is made of metal. Where it is necessary to have a one-way ploughing, the turnwrest type of ploughs are used. Lakhs of such ploughs are in use in Maharashtra.

There, the ploughs are now available on hire also. It is being adopted where green-manuring crops or surface vegetation have to be buried to hasten decomposition. Heavy mouldboard ploughs require sturdier animals, and sometimes two or three pairs of bullocks are used. They are also employed for the eradication of deep-rooted weeds, e.g. *kans (Saccharum spontaneum)* and *kunds (Cyperus rotundus)*. The turnwrest type of mouldboard ploughs are very efficient. For lighter ploughing, the beam-type ploughs are used. The shares of the lighter ploughs should be made of steel instead of cast iron. These ploughs are finding increased favour in several parts of northern India and are replacing the *desi* plough. The mouldboard plough in common use is with bullock-power cut furrow slices varying from 12 to 23 cm in width. Typical figures in respect of their relative performances are given below:

Double Mould Boards: These ploughs have double mouldboards. There are three sizes—the light, the medium and the heavy. The heavy ones are used for the cultivation of sugarcane and the lighter ones for the cultivation of vegetables and other crops. Some of the improved ridger ploughs have an arrangement for adjusting the width of the furrow by changing the distance between the two mould boards. For vegetable orchards, and for farmers who grow sugarcane, potatoes, etc., the possession of the ridger is a must.

Width of the plough (cm)	-	17.78	16.31 (nominal)	10.43 (nominal)
Width of the furrow cut (cm)	-	17.78	13.34 to 16 31	9.16 to 10 43
Depth of the cut (cm)	-	12.7	12.7	12.7

Contd...

Drought in dry sandy loam				
at Delhi (kg)	-	123.37	127	118
Drought per square inch of				
furrow cross-section (kg)	-	0.562	0.632	0.913
Speed of bullocks (km)		3	1.673	1.737

Harrows and Cultivators: The indigenous harrows and cultivators are known as *bakhars* and *guntakas.* They mostly consist of a long sharp blade fixed to a wooden implement. The body of such an implement is made of *babool* wood, cut into a rectangular or octagonal shape. The length of the body is 91 to 182 cm and the width is 25 to 30 cm. The blade is attached to the log on its underside with two slanting wooden pegs. These implements are used for preparing the seedbed, for covering the seed and also for summer cultivation, provided the soil is not too hard.

Improved *bakhars* and *guntakas* have also been, lately, manufactured. The improvements relate to the changing of the blades, the adjustment of the depth and the attachments that may be needed in the case of these harrows. The length of the blade is from 61 to 152 cm and, therefore, this implement can cover quite a large area in a day. Steel *bakhars* and *guntakas* are nowadays mass-manufactured in India. Their prices range from Rs 40 to 55.

Some of the harrows used in foreign countries, e.g. the spike-tooth harrow, peg-tooth harrow, the spring-tooth harrow, and disc harrows have also been introduced. Out of all these harrows, the disc harrows have proved to be very popular. The spike-tooth harrow is also often used, particularly in parts of Punjab and western Uttar Pradesh. The spike-tooth harrow is also often used for inter-cultivation and for thinning crops in Punjab. The triangular frame on which spikes are fixed is used in that state.

Different Processes

Soil Implements

Soil for Scoop: A soil scoop consists of a shallow box-like structure with an open front. On its front, a sharp blade is fixed and two hooks are attached to it. The bullocks are yoked to the hooks. Two wooden handles are provided at the back. The capacity of the soil scoop is 0.056 to 0.084 cu m. This implement is used when the soil is to be carried to a fairly long distance. The price of the soil scoop is Rs 60 to 1,000.

Wheel Barrows: Common wheel barrows are very useful for gardens or orchards. The wheels may be of mild steel or of rubber. The price of a wheel barrow is Rs 150 to 200.

Keni Making or Levelling Karaha: It consists of a fiat blade 1 to 5 m long, with a vertical handle and two rings in the front for yoking the bullocks with ropes. This is probably the commonest implement used in soil conservation for making bunds and levelling the land. Literally, thousands of these implements are used in Ahmadnagar district of Maharashtra for bunding. The bunding of about 200,000 ha has been completed with this implement. It is also becoming popular in other parts of Maharashtra, Andhra Pradesh and Karnataka for bunding land as well as for bunding *nullahs.*

The space where the bund is to be formed is first marked, and on both sides of this space, a width of about 5 m is ploughed with a mouldboard plough. The loosened soil is then moved on to form a bund. If the land to be ploughed is hard and cloddy, it hurts the feet of the bullocks and reduces their life. As far as possible, the bunds should be formed when the soil is soft. The price of the *keni* is Rs 40 to 60.

Patela: The indigenous implement is called *patela* or *sohaga.* It is merely a flat log 1.8 to 2 m long and is dragged over the clods by bullocks. The driver usually stands on the log to put weight on it and hold it down.

In Uttar Pradesh, the improved *pata* or *patela* is now mass-manufactured and has become quite popular. It is called *Singh patela.* It is provided with a series of pointed hooks made of 10 mm to 15 mm rods of mild steel. These hooks help to crush the clods and to collect the weeds. This implement costs 100 to 150 rupees, depending upon the length of the wooden blade and the number of steel rods used.

A land-leveller or *jandra* is also used in some tracts. It consists of a plank of wood, about 1 to 1.5 m long, 12 cm and 2.5 cm thick. In those areas where bamboo is readily available, a ladder-shaped leveller made of bamboo is used.

Bunds Forming: For soil conservation, sometimes it is necessary to make temporary low-level bunds, 18 to 22 cm high. These bunds last for one agricultural season. They either get washed away or are obliterated by subsequent ploughing and cultivation. For forming such bunds, an implement, called bund-former, was originally developed at Coimbatore. The implement consists of two concave blades, 90 to 150 cm long and 10 to 15 cm wide, joined across in a conversing manner. This implement is to be worked in a well-ploughed and harrowed field, just for collecting the soil and not for digging it. Its job is to collect the soil at the wider front and deposit it in the form of a ridge or a low bund behind. There are three sizes of bund-formers with blades 9 and 12 cm long. Recently, a bigger bund-former, called "sara-making" machine, has also been designed. A wooden bund-former that would cost less than 10 rupees has been prepared at the Agricultural Workshop at Nilokheri (Haryana). Its blades are made of *babool* wood and its construction is so simple that the implement can be easily made by a village carpenter. If the wood is made available by the farmer, the implement would not cost more than 6 rupees.

The Forest Department in Madras has developed basin-listers for forming a series of a small basin or depressions in the land in order to collect rain-water. This is a useful operation in dry farming, because it conserves moisture. However, the implement, being rather costly, has not become popular.

Seed-sowing: Among the indigenous implements, many seed-sowing devices are attached to implements, such as the *desi* plough. They are commonly called *nari* plough, *tifan, argada* and other multi-type implements that are used in the Deccan. Tube or tubes are fixed to these implements and on the top of these, a seeding-bowl, in which the seed is dropped, is fixed. The seeding-bowl has a number of holes in it. The construction of the holes is ingenious, with the result that the seed is dropped fairly evenly into the tubes and goes down to the tubes. In some areas, which are considered backward agriculturally, broadcasting or the sowing of seed at random on the cultivated field is also followed but the major portion of the area is nowadays sown in line and the seed is drilled. Sowing in lines ensures the proper distribution and placing of seed, resulting in the uniform spacing of the plants, with a concomitant reduction in seed-rate. It also helps the seeds to germinate regularly, ensuring a better stand of the crop and also facilitates inter-culture with bullock-operated hoes, thus reducing the cost of cultivation.

Quite a large number of improved seed-drills, both automatic and hand-operated, have been designed. Among the hand-operated ones, a copy of the wooden *tifan* has been made in Uttar Pradesh. It is made of mild steel, with an arrangement for changing the distance between the tines. The price of such a seed-drill is Rs 75 to 100. Among the automatic seed-drills mention may be made of the McCormic seed-drill, the Master seed-drill, the Gunti seed-drill and the Kirloskar seed-drill. The McCormic seed-drill uses an external-force feed-type of seed-metering device. The price of this seed-drill for a three-tined implement is Rs 1,200. The Master seed-drill has a series of small cups or spoons, fixed on the discs. These discs revolve in a box filled with seed. The spoons gather the seeds and drop them into the tubes. The price of the Master seed-drill is Rs 950. The Gunti seed-drill also uses an external-force feed-type seed-metering device, but it has also a fertilizer-metering box fixed alongside the seed-box. The Kirloskar seed-drill which has recently been introduced has an ingenious device for seeding but it has not become popular. Most of the manufacturers of seed-drills are now making efforts to combine the simultaneous feeding of fertilizer with the dropping of the seed, thus making the implement a seed cum fertilizer drill. So far, however, none of these combined devices has proved very popular because of high cost, complicated designs greater draft, etc.

The seed-drills need a lot of skill in operation. It is necessary, particularly in respect of automatic seed-drills to calibrate them before sowing, so as to know the exact seed-rate that it will drop. The seed-drills are to be calibrated for each type of crop separately because of the differences in the shape, size and weight of the seeds of different crops. Similarly, in the case of the automatic seed-drills, proper care has to be exercised in choosing the soil-working part, such as the tine. If the soil is hard, strong straight tines with sharp edges in front should be selected. If the soil is soft, the disc shovels may be preferred. Some other important factors which a designer of a seed-drill and its user have to keep in mind can also be mentioned. These are the covering of the seeds, the pressing of the soil slightly over the seeds, the transport facilities for these drills, etc. Among the sowing devices, mention can also be made of transplanting-machines. Even though strenuous efforts are being made in different countries, such as Japan, Italy, the USA and China, so far no really useful machine for transplanting paddy has been prepared, except in Japan. Some of these new machines have been brought for trial at the Central Tractor Testing and Training Station, Budni, and have given quite good results as far as transplanting is concerned, but the machines require specially raised seedlings in trays and this is an elaborate and costly process. Therefore, transplanting has to be done by hand. A small tool for transplanting two or three seedlings of paddy at a time is being used in Burma and Malaysia. There is need to introduce this tool in India, because with little practice it can increase the efficiency of the labourers at least by 20 per cent.

Most of the interculture in India is done with a hand-tool, called *khurpi*, which consists of a sharp-edged triangular or sickle-shaped blade. Sometimes, weeding is done with a small spade, hoe or *mummuty*. Most of these tools have small handles and the user has to sit on his heel? or bend his back while weeding.

Weeding is one single operation which can increase crop yields without much additional investment, because efficient weeding will save moisture, crop nutrients and, to some extent, air and light required for the cultivated crop. To increase the efficiency of weeding, long-handled hoes, such as the Singh Hand-hoe, the Sharma Hand-hoe and the Dutch hoe have been introduced. The price of the hand-hoe is around Rs 6, including the handle and it is well within the means of even a poor farmer. It has been observed that four to five times more of weeding can be done with such long-handled hoes than with a *khurpi*. These hand-hoes are simple in construction and can be made by the village blacksmith. The use of these hand-hoes is, therefore, strongly recommended.

The wheel-hoes are a further improvement upon the hand-hoes. A wheel-hoe consists of a body, handle and a wheel. To its body, different attachments, such as shovels, a small plough, a small ridger, and other types of blades can be fixed. The price of the wheel-hoe is Rs 60 to 75, with a Weeding-attachment. Other attachments can be purchased at extra cost. Being provided with a wheel, it is easier to work and is very useful for inter cultivation in vegetable gardens and orchards. Where the soil is hard, one man can pull this hoe and another can direct it. Where the soil is soft, it requires only one man to work it.

For hoeing line-sown crops with bullock power a light *desi* plough is used in northern India. In the area between the Narmada river and the Krishna river, multi-tined bullock-operated hoes are used. They are known as *doura (Gunti), douri, dundia* and *dulari*. These are the miniature forms of *bakhars*. When the plants are small, multi-tined hoes fixed to a single wooden body are also used in Andhra Pradesh and northern Karnataka. The draft of these hoes is very light and sometimes two to four of them are pulled by one pair of bullocks. In the dry farming areas of Maharashtra, a *doura* with a split blade is used. The blades in this case work on each side of the row of plants. In the case of rain-fed crops, hoeing by using bullocks is done at the most three or four times, after which the plants grow tall and bushy and also the soil becomes hard. To cope with this condition, stronger and sharp hoes have been designed. They are commonly known as the Washim hoe, the Akola hoe, the Baroda hoe and the *triphali*. The designs of these hoes are more or less similar. They consist of three tines with a sharp triangular blade fixed to each. These hoes are very useful for the interculture of the cotton crop sown in rows.

Yet another implement called Indore Ridger, was introduced in Malwa some years back. This implement has a sharp beam, with a sloping ends and a hexagonal share attached to it with a strong iron bar. The ridger can be used up to a later stage of the crop without doing physical injury to the plants. It not only eradicates weeds but also produces a good mulch. Similar to the *doura,* two or three such ridgers can be attached to a pair of bullocks. It is surprising that such bullock-drawn inter-cultural implements have not yet been introduced into northern India despite their being used by the Indian farmers in the Deccan for centuries. There is a great deal of scope to introduce such implements into those areas where they are not being used at present.

For some special crops, such as paddy, special types of paddy-weeders have been designed and are being mass-manufactured in India. They are based on the Japanese design. They consist of two rollers with pointed fingers and are pushed in between the rows of the paddy plants when water is standing in the field. The idea of using these weeders is not to uproot the weeds and bring them to surface, as in the case of other weeders, but to bury them in the mud so as to suffocate and decompose them, and add organic matter to the soil.

With the advancement of scientific research, new methods to control weeds with selective weedicides, such as 2, 4-D, are being used. When this weedicide is sprayed in between the rows of a crop it kills the weeds only and not the crop. Sprayers are used for spraying these weedicides. At present, the weedicides are costly, but when their supply becomes abundant, the killing of the weeds with selective weedicides may become quite popular in India and it may not be necessary to use other implements to a great extent.

The other implements may have to be used for forming a fine mulch in between the rows, so as to conserve soil moisture. Some of the intercultural implements may have some attachments fixed to them to carry out some special jobs, such as sulphur-dusting in tobacco cultivation and the top-dressing of fertilizers. Already, simple sulphur dusters which can be worked in rows, and fertilizer attachments to the weeders have been designed.

Harvesting: So far, the harvesting operation has not yielded to any mechanized tool. Almost universally, sickle is the only tool that is used for harvesting crops in India even though efforts have been made to introduce reapers, mowers, and even combined harvesters and threshers. These machines have, however, proved successful only on big estates or government farms, but because of their high cost, they have not become popular with the Indian farmers except when used on custom basis, as in Punjab and in Haryana during the last five years. The European scythe has also been tried, but except for cutting of lucerne or some such green fodder they have not proved very useful. One of the main reasons seems to be the shattering of the grain because of the impact of the harvesting implements and machines. In the case of sickles, there is less of shattering. The sickles are of two types, namely, plain or serrated or saw-shaped. The serrated sickles are used for cutting paddy and wheat.

Thrashing: Threshing in, the case of paddy, if the area is small, is done mostly by beating the harvested plants with sticks. In the case of other crops, it is done by treading the crop with a team of animals. Stone rollers, about 90 cm in length and 30 to 60 cm in diameter, have also been used in some areas. However, no other thresher has become so popular as the "Olpad Thresher", which consists of 14 to 21 serrated discs fixed on a rectangular wooden frame in three lines. A seat is provided for the driver and also a guard. The implement is pulled by a pair of bullocks and worked on the harvested material which is spread-thickly on the threshing-floor. Threshing with this thresher is not only cheaper but is also quicker and more efficient than treading by bullocks and by using other methods.

The Japanese pedal-operated thresher, which was introduced into the rice-growing areas has not proved very popular because it does not give as much yield of grain as beating with sticks does. The reason for the Japanese using this type of thresher is that they are keen to get as long pieces of straw as possible in order to prepare ropes, mats or sacks out of the paddy straw. These articles are not used by the Indian farmers and, therefore, this thresher has not become popular in India. It can thresh about 300 kg of paddy in a working day of eight hours.

The threshing of wheat in some of the large private and government farms is done by trampling the harvested crop under a tractor or under a disc harrow attached to a tractor. It has been noticed that a 15-bhp tractor with a suitable harrow produces about 1,860 to 2,975 kg of clean undamaged grain per day of 10 hours at a working cost of Rs 35 to 40. The *bhusa* or the broken straw is also of good quality.

Recently, the Government of Punjab has introduced a power-operated wheat-thresher. It is worked with a tractor or an oil-engine of 20 to 25 bhp. This thresher is being simplified so that it can also be operated with an oil-engine of 10 to 15 bhp. The thresher is used by the owner alone if he himself is a big farmer, or can be given to other farmer, or can be given to other farmers on custom basis, or can be used co-operatively. In most of the villages in Punjab, Haryana, and western Uttar Pradesh, now at least one thresher exists, and in many cases there is more than one. Thousands are used on custom basis. The *bhusa* is a precious commodity and finds a ready market as a cattle feed. Therefore, a thresher, which cannot make *bhusa, w*ill not become popular. At the research stations in the country, the Japanese power-operated paddy-thresher is being given a trial. This machine can turn out 50 kg of clean paddy per hour. However, it has no *bhusa-making* arrangement. To make *bhusa,* the straw has again to be trampled under the feet of the bullocks. The drum-type power-operated thresher of 10 to 15 bhp has become very popular in northern India. This thresher is an example of a machine which the farmers in that region have readily adopted because of its great utility. Now virtually threshing by trampling straw by using, bullocks is limited to very small farmers. The introduction of this machine has saved much labour and has made threshing very easy and to complete it before the monsoon.

Different Winnowing-fans and Winnower Tools: To separate the chaff from the grain, the threshed material is put before a good natural breeze and the grain is allowed to drop through gravity and be separated from *the bhusa,* which being lighter, falls a little further away from the grain heap. In many areas, however, the natural breeze at the time of winnowing is not readily available and the farmers have to wait for hours and hours even for days to avail themselves of the natural breeze. Usually, in such areas, there is a danger of rain and hail-storm during this period. The farmers are, therefore, very anxious to take the season's produce to their homes as early as possible. Some crude methods of producing artificial breeze have been used by the farmers, but recently, winnowing-fans, similar to the ceiling fans, have been introduced. These fans may be operated by hand or with a pedal and the number of revolutions are increased by using a pair of gears or with sprocket and chain or a belt. In some cases, e.g. the Nilokheri winnowing-fan, such a drive is not necessary. The size of the fan itself is such that it can produce a good breeze. The object of these fans is to produce a breeze only and, therefore, no sieves are provided.

The winnowers are the machines with a series of three to four sieves to do the preliminary grading. A fan is enclosed in a box-like arrangement and is rotated fast. The sieves are also provided with a shaking arrangement. The steel winnowers cost Rs 1,000 to 1,500 whereas the wooden ones cost Rs 800 to 1,000.

Machines Tools Processing: Under this heading, the machines that process the farm produce during the post-harvest period have been described. These machines include sugarcane-crushers, fodder-cutters, groundnut-decorticators, paddy-hullers or paddy-shellers.

Crushers for Sugarcane: The growers of sugarcane cannot dispose of their entire harvest as cane. Part of it has to be converted into *gur* or jaggery. *Gar* is also a popular article of food in the villages. Formerly, the juice was extracted from the cane with crude two-roller wooden crushers. With the growing of improved varieties of cane, a three-roller cast-iron crusher operated either by bullock power or by engine power has been put into the market. It has greater crushing efficiency and, therefore, farmers are either buying such crushers or taking them on rent during the season.

Cutters for Fodder: Of all the improved agricultural implements, the fodder-cutter has become most popular, particularly in Punjab, Haryana, Delhi and Uttar Pradesh areas where green *jouar* or corn is required to be cut for feeding to the cattle. Two sharp steel blades are fixed to a circular flywheel. An automatic pushing arrangement is also provided. In some areas, where power is available, farmers are using the power chaff-cutters or fodder-cutters, or getting their chaff cut on rent basis. Chaff-cutters with good-quality blades are now manufactured in large numbers in India and are available at fairly reasonable prices. The power-driven chaff-cutters can be used also for cutting dry *jowar* stalks, which is not easy to cut with a hand operated chaff-cutter. For farmers, who own small holdings, the *Gopuri* type chaff-cutter designed at Sewagram near Wardha is recommended. It costs only Rs 40 to 50, whereas the cost of the hand-operated circular chaff-cutter is Rs 200 to 300.

Decorticators for Groundnut: When groundnut is to be sown, women labourers are employed for breaking the pods and taking out the kernels. In doing so, the testa of the kernel also gets separated to a considerable extent making it useless for sowing, because such damaged kernels will not germinate. Groundnut-decorticators, therefore, have been designed. The Maganwadi type of groundnut-decorticator is like a cradle with a screen placed below. The other rotary decorticators, either hand-operated or power-operated, are also available. The price of a hand-operated decorticator is Rs 200 to 300 arid the price of a power-operated one is Rs 600 to 1000

Hullers for Paddy: Paddy-hulling or paddy-shelling has been done in India with a *dhenkli* or wooden lever. The Khadi and Village Industries Commission has introduced a Rajasthan type of huller which is quite efficient. Recently, a manufacturer in India has put into market hand-operated hullers similar to Japanese types. They cost Rs 1,000 to 1,500, but are good for small holdings. They are operated by two persons with levered handles. The shelling or hulling is accomplished by the impact of the grain on a very fast-revolving rubber-disc.

A manufacturer has also put on the market a paddy-polishing machine. After the paddy is shelled or hulled, the grain is polished with this machine.

Other Machines and Tools

Apart from the manual labour, the greatest amount of farm power in India comes from animals. Among the animals, bullocks are used most. The efficiency of using bullocks depends upon their feeding, maintenance, manner of yoking and training. The draught exerted by a work animal is related to its body-weight, the mechanics of working, the method of hitching the implement and the type of harness or yoke used. The following table shows the range of the ratio between the weights of the bullocks and the draught exerted in ploughing.

Body Weights of Bollocks and the Draught Exerted in Ploughing:

Average body-weight per pair of bullocks (kg)	*Locality*	*Draught (kg)*	*Weight draught ratio*
432	Bankura	66.678	6.46
463.592	Tripura	46.449	10.0
453.592	Patna	53.9795	8.33
726.747	Jhansi	59.874	11.95
771.107	Delhi	66.178	11.60
771.107	Aligarh	78.471	9.82
771.107	Coimbatore	146.149	5.20
907.194	Baroda	138.616	6.55
952.553	Karnal	134.535	7.07
1166.670	Indore	183.064	6.00

It may be assumed that the bullocks exert roughly 1/12 to 1/5 of their body-weights as draught. There is reason to believe that the draught animals generally have the ability to exert at least for a short period much greater pull than that indicated in the above table. There are cases on record, where a pair of bullocks weighing 725 to 900 kg has worked mouldboard ploughs having a draught ranging from 145 to 181 kg continually for 30 to 40 days without any adverse effect on their health. That is why sometimes a pair of good bullocks can do a better job than a 4- or 5-hp small tractor, which requires a very considerable amount of its horse power for its own movement. It cannot exert even for a small period appreciably increased draught which a pair of bullocks can exert.

Bullocks, buffaloes, or camels are commonly used in India as work animals. Camels are mostly used in desert areas and also for Persian wheels, oil *ghanis* or *kohlus* and to transport carts in Punjab, Rajasthan and U.P. Since bullocks are far commoner than other animals, some work has been done in determining their horse power. Their power is related to their body-weights. The following table shows the average body-weights of mature bullocks of the main Indian breeds used in farm operations:

Breed	*Average body-weight (kg)*
Nimari	303.907
Nagori	317.515
Siri	317.515
Sahiwal	408.233
Gaolao	408.233
Tharparkar	430.913
Kangayam,	522.601
Deoni	522.601
Haryana	508.023
Krishna Valley	544.310
Kankrei	544.310

Some tests have indicated that the horse power of two bullocks operating a Persian wheel is 0.877 and of those engaged in ploughing varies from 0.29 to 1.87. The walking speed of bullocks working in a circle with a radius of 3 to 4 m, while operating a Persian wheel is 2.5 miles per hour, with a draught of 50 to 51 kg. Bullocks engaged in ploughing walk at a speed of about 1.7 miles per hour and slow down to 1.2 miles per hour, at the end of the day's work. Thus at this speed, a pair of bullocks will plough 0.8 to 0.12 ha in a day with a *desi* plough and 0.2 to 0.4 ha in a day with a medium type of mouldboard plough in an alluvial soil.

Improvement in Yokes and Harnesses: It has long been realised that with improvement in the harness, the yoke and the method of hitching, the implements are likely to lead to a more efficient utilization of draught. Experiments in this connection are going on at Coimbatore, Allahabad, and Bardoli and a number of other places. The experiments conducted at

Allahabad have shown that among the indigenous yokes, the Nagpuri yoke gives the greatest draught. This yoke has now been introduced, at least on an experimental basis, almost all over India and is being manufactured by some of the firms in U.P. The yoke is made of *babool* or *kikar* wood. Because of its shape and curvature, it rests properly on the necks of the bullocks and thus gives a greater draught with comfort to the animal.

In case a good yoke is not used, the animal gets a neck gall and thereby reduces its efficiency. The weight of the yokes used in India varies from 7.25 to 15.875 kg depending on the type of wood used and the dimensions compatible with the strength required to withstand the pull exerted by the bullocks. However, most of the two-bullock yokes are good for a pull of 45.359 to 54.431 kg which is the draught commonly required for agricultural implements in use, particularly of the indigenous types. Efforts have also been made to introduce one-bullock yokes, but not with success.

Conservation Soil and Water of Irrigation and Drainage : Soil and water conservation go hand in hand. It is necessary to make the best use of the available rainfall in the scarcity areas, particularly in the districts, like Ahmadnagar, Rayalaseema and Sholapur. Some of the implements which are used for moving the soil or for making bunds have already been described. These implements include the soil scoop, the bund-formers, *keni,* etc. Their description, therefore, is not repeated here.

The dry-farming implements, such as a long-bladed *bakhar,* a seed-drill having a distance of 3.81 cm between the tines and specially designed intercultural implements are recommended for dry-farming operations, the principal idea being to utilize, as far as possible, the largest amount of rain water. By bunding the fields and also the *nallahs,* some quantity of water is absorbed and conserved by the soil. This water, in turn, accumulates in open wells or tanks and can then be utilized for irrigation.

Another set of implements and machinery used for preparing the land for irrigation consists of a float, a drag, and a leveller. The float consists of a rectangular wooden frame to which cross plates are fixed. The last cross plate is inclined in the direction of the movement of the float. The float is mostly used where the land has already been levelled fairly well and where it is necessary to level it further in order that the irrigation water flows over the land uniformly and does not accumulate in depressions. A drag is an implement which throws the soil in one direction, making the channels flat-bottomed.

The leveller is similar to the *keni,* but is longer for the uniform levelling of seedbeds. Water is carried through irrigation channels from wells to the fields and in order to minimize the loss of water through seepage, it is necessary to line them either with bricks or with cement. Recently, the workers of the Indian Road Research Institute have made a cheaper type of chemical out of creosote. This material can be used to plaster earthen water-channels. A kind of weedicide is also mixed with it to prevent weeds from growing in the channels.

One of the major difficulties of the farmers is the raising of water from the wells, tanks or canals to irrigate the fields. Since time immemorial, various indigenous devices have been used for lifting water, the simplest ones being a *dhenkli,* a Persian wheel, a *mhot,* a *pikota,* and *dons.* Where the water level is just near the ground, swing baskets are used. They can lift water from depth of 0.9 to 1.2 m. Two persons are required to work this swing basket. For lifts up to 1.828 m boat-shaped wooden or metal scoops, called *dons,* are used in Orissa, Bihar and Uttar Pradesh. The *dhenkli* consists of a lever and a bucket suitable for lifting the water from wells Up to 4.572 m from the surface of the ground. In the case of a *pikota,* which is common in Tamil Nadu, the operator has to walk on a levered beam. The capacity of the bucket of the *dhenkli* varies from 9 to 18 litres and its water duty, as investigated in Uttar Pradesh is less than a hectare.

The leather *mhot or charsa* is common in Bihar, Uttar Pradesh, Rajasthan, Madhya Pradesh and Maharashtra. The *mhots* in northern India are merely big leather bags, whereas the *mhots* in Andhra Pradesh, Maharashtra and Karnataka have a spout which allows it to be emptied automatically. They are worked by one pair of bullocks and up to depths of 9 to 12 m, but when the lift is more than 15 m, as in Rajasthan, two pairs of bullocks are used.

The capacity of the bucket varies from 136 to 181 litres depending on the depth of the well. The discharge is about 6,819 to 9,092 litres per hour and it takes three to four days to irrigate 0.404 hectare. Some manufacturers have also prepared sheet metal *mhots* differing in design and capacity and incorporating into them small improvements like automatic valves in the bottom, self-balancing paired buckets or the bullocks operating in a circle instead of on a slope. The discharge per hour of the improved *mhots* is almost the same as that of the leather *mhots* used by the farmers.

The Persian wheels, or *rakats*, are extremely common in northern India, Uttar Pradesh, Rajasthan, Punjab and northern Maharashtra. Wherever the water level is between 4.5 and 9 m below the ground level, they are popular. A Persian-wheel consists of an endless chain and mild steel flats to which are fixed galvanized-iron buckets. The endless chain rotates on a large cage wheel. The axle of the cage wheel is geared with a pinion to the big gear, which is rotated by a pair of bullocks or a camel walking in a circular track. The modern Persian-wheels are fitted with ball-bearings, and several improved types have been produced. Mention may be made of 'Mayadas' double Persian-wheel which has found some favour in Uttar Pradesh. The average draft of a Persian wheel working up to a depth of about 4 m comes to 59 kg and that working up to a depth of 7.3 m, 86.1 kg. The discharge ranges from 8184.3 to 22730.4 litres per hour and the efficiency ranges from 34 to 44 per cent.

With the advent of engine-power or electric-power, a very large number of Indian farmers are going in for centrifugal pumps. This tendency is invariably growing where there are cash crops, or orchards or gardens. The centrifugal pumps are worked with internal-combustion oil-engines or electric motors. The state departments of agriculture also give *taccavi* loans, and other facilities for the installation of centrifugal pumps. In operating centrifugal pumps, priming is very important, because the centrifugal pump works on the principle of vacuum. The seals of the centrifugal pump should also be tight. In the case of those wells where the water level fluctuates considerably between the rainy season and the summer season and where it is necessary to apply irrigation in summer, additional platforms have to be installed inside the well for fixing the pump at different levels. Because a centrifugal pump has the limitation of raising the water up to the maximum level of 7 m theoretically and up to 5 m practically from the level of the water, the centrifugal pump has to be fixed at a height of 5 m from the water level to get the best out of it.

With the assistance of some of the foreign firms, the Indian manufacturers are also putting into market submersible or floating pumps that rise or go down according to the fluctuating depth of the well. In such cases, it is not necessary to put one or two different plates for fixing the pumps. For very deep wells, the turbine pumps are recommended. These are manufactured specially on order. For tube-wells also, the turbine pumps are useful. In the case of filter-point tube-wells in the coastal region of Andhra Pradesh and Tamil Nadu, the centrifugal pumps can be used, because there the water level is, on an average, 5 m below the ground level. The filter-point tubewells fitted with Such centrifugal pumps are very cheap and, therefore, their use is becoming extremely popular in these regions.

The use of a hydraulic ram in the hilly areas is also recommended. It does not require any electricity or mechanical power. The energy that is developed by the falling of the water to a depth 2.438 m is used to raise it to the field level. Such hydraulic rams can be conveniently used in the hilly areas, such as Himachal Pradesh, Manipur, Mizoram and the Western

Ghats. Irrigation water, even if it is readily available, should be judiciously used ; otherwise, it not only damages the crops but also damages the soil to the extent of making it barren. The injudicious use of irrigation water raises the salts in the soil to the surface, making the growth of plants impossible. It is only by the addition of organic matter or of gypsum or both that it is possible to reclaim such lands. Therefore, the extravagant use of irrigation water should be avoided. In those soils where water often accumulates, it is necessary to provide for drainage.

Drainage Machines: Good drainage is complementary to irrigation. In those areas where there is plenty of irrigation water, as in Punjab, Haryana and western Uttar Pradesh, drainage has become almost a national problem and is agitating the minds of the farmers and of the government. It is, therefore, necessary to know the methods developed for draining away the excess water from the fields. Water accumulates in the soil either from rains or through seepage from canals or irrigation channels and reservoirs.

The high subsoil water-table damages agricultural land in many ways, leading particularly to the accumulation of undesirable salts in the root-zone or on the surface of the soil, involving drainage problems. The root cause of the drainage must be thoroughly investigated and understood. The problem of persistent water-logging is often complex and expert advice might be sought to solve it. In areas of heavy rainfall, good surface drainage has to be arranged. The drainage ditches should be wide and of trapezoidal or parabolic shape, so that they offer a rapidly increasing cross-section to the flow of the stream as the water level in them rises. On relatively flat land, they may be made to run down the slope. The velocity of water in the drains should be maintained at non-erosive level by proper grading and shaping. The spacing of the field drains should be worked out taking into consideration the character of the soil, the slope, the intensity of the rainfall, etc. A field drain should be laid to join a utility channel of adequate dimensions. These utility channels should be sufficiently big to take the maximum quantity of water.

In those soils where water accumulates underground within a depth of 0.6 to 1.2 m, underground drainage is needed. In case this high water-table is due to percolation from an external source, intersection drains should be installed to catch and divert the water. There are several types of underground drainage systems, but most of them are costly. As the work is of a permanent nature, it involves considerable expense. A master plan for draining the land should be prepared in consultation with specialist engineer and after making a preliminary survey of the levels. The best kind of drainage pipes are hard, burnt, earthenware flange-less pipes. The diameter of the pipe is usually 10 cm. The pipes should be laid well below the lowest root level of the crops proposed to be grown on the land. The joints of the pipes are important and should be left uncemented, but protected with a strip of tarred paper to prevent the entry of silt or soil, as the drainage of water takes place largely through the joints and not through the porous body as is usually thought. The spacing of the drains depends on the level of the land and the permeability of the soil. The drains can be placed in a grid or herring-bone fashion, depending upon the level of the water-logged area. In the case of sugarcane-growing areas in Andhra Pradesh and Tamil Nadu, the farmers are using mole drains. They are made at a depth of 0.45 to 0.6 m and are quite effective. The tractor-mounted trenching-machines are very useful in digging the channels for putting the tile drains.

In order to drain the water from the fields, the Japanese farmers use vertical pumps having low lifts, and high volume. Samples of such pumps were recently obtained from Japan and suitable designs were prepared. Such pumps are now available in India. The capacity of such pumps is 136,380 to 181,540 litres per hour and they can be effectively used for draining the soil, if the utility channels are available.

Tractors: As explained earlier, even though we have to lay greater emphasis on manually-operated and bullock-drawn implements and machinery for the present, the slow mechanization of agriculture with power-operated implements has begun to come of age. It is generally considered that one of the main reasons for high agricultural yield in the Western countries is a high degree of mechanization of various farm operations. Mechanisation is possible there because of the shortage of labour and the overall development of engineering industries. So far, these two conditions have not existed in India. The labour has been relatively cheap and the industries are as yet, more or less, underdeveloped. However, these conditions have been rapidly changing since Independence. Agricultural labour, particularly at sowing and harvesting of crops is getting scarce. Almost any kind of agricultural implement or machine can now be manufactured in this country and that tool of good quality and at a reasonable price. The farmers are becoming aware of efficiently running a farm as a business. They are calculating the inputs and the resultant production, and agriculture is no longer a way of subsistence living. The governments of the states and the Centre are also doing their best to bring home to the farmers the results of research. These changing conditions necessitate the going in for mechanization and a fairly rapid progress is being made. That mechanization can succeed has also been shown by the farmers having already adopted diesel-engine pumping-sets as well as electrically-driven water-lifts. In certain states, quite a large number of farmers are using tractors and tractor-drawn implements. It may not be possible to substitute tractors for bullock power completely, but for the following operations, mechanized power can be profitably and effectively used and needs to be encouraged:

(1) For carrying out heavy jobs, such as summer ploughing or for reclaiming vast stretches of land. This work cannot be done either with manually-operated or bullock-drawn implements, as it needs lot of power. Many state governments have their own tractor units, which the farmers take on hire for deep ploughing, for reclaiming land and for carrying out soil and water-conservation

practices, such as the bunding, levelling and terracing of the land and for good water management

(2) for eradicating weeds like *kans, hariali, dub* etc., which are deep-rooted and obnoxious ;

(3) for clearing jungles and for opening new lands ;

(4) for the desilting of tanks ;

(5) for big government farms, tea or coffee estates, and for using them on collective basis ; and

(6) for custom-hiring, wherever possible.

All the above types of operations have been done mainly through government agencies in the past. The best example is the Central Tractor Organisation of the Government of India, which functioned from 1947 till some years ago and reclaimed lakhs of acres in central India. Such large scale operations require heavy crawler-type tractors and specialized equipment, which are neither normally required by individual farmers, nor are they within their means.

Necessity of Tractor: For attachment to the tractor, we must provide the farmers with different types of implements to carry out different agricultural operations. Because of the higher horse power used, the tractors can do many more jobs which cannot be done by using bullock power. The construction of a tractor is similar to that of a motor-car in some ways, with the difference that the usual fuel used is diesel, the wheels are bigger and the speed slower.

The construction of a tractor is sturdy, so that it can work in rough fields and arrangements have to be made for attaching different implements to it as well as for running stationary machines by using the power takeoff. The other phases of construction are similar to those of a motor car, such as the engine, the clutch plates, the transmission gears, the differential, etc. Thus, one can see that a tractor is a modification of a motor-car to perform agricultural operations including bund-making, etc. The tractors can be classified into the following types:

Crawler Tractor: Such tractors are suitable for relatively high-draft and low-speed work. They can also be used on uneven ground or on soft marshy land where the wheels would not work. These tractors are heavier but their weight is more evenly distributed on a bigger area ; hence they do not compact the ground as much as the wheel-type tractors do. Their horse power ranges from 60 to even 500. They are employed for reclaiming large pieces of making the barren land fit for raising crops, for desilting tanks, for preparing barriers or bunds, for levelling the land and for draining away excess water.

Four-wheeled Tractors: Under this category come most of the agricultural tractors with an average horse power of 30 to 35. These tractors are fitted with pneumatic tyres. They can be used for ploughing, harrowing, sowing, harvesting, transport and belt work. They can also be used for intercultivation by making use of an arrangement by which the width between the wheels can be changed. This arrangement enables the tractor to be manoeuvred between the rows of growing crops. By using trailers, the tractors can be used for transporting manure, seeds, fertilizer, etc. These tractors are also being used for belt work such as for cutting chaff, lifting water and for working a wheat-thresher.

Small Size Tractors and Powar-tillers: A small tractor can be defined as machine having four wheels or two axles. A power-tiller is a machine having two wheels or one axle. The term "power-tiller" originated in Japan and most of the machines there are fitted with a rotary implement behind such a single-axles, two-wheeled small tractor. The horse power of the small machine ranges from 3 to 15. The Indian farmers are probably showing the greatest interest in such small tractors and power-tillers for the simple reason that the price of a good pair of bullocks is increasing alarmingly and that the bullocks require a lot of expense for their maintenance. The Indian farmers are looking forward to using small tractors or power-tillers in the range of 5 to 10 hp, which would not cost more than its 2,000 to 3,000. The main jobs that such machines should be able to perform are ploughing, harrowing, puddling, transport, pumping water and spraying insecticides. If the machines within the range mentioned above can do these jobs and are available for the price mentioned, they would certainly prove to be very popular. Such small tractors and power-tillers would also prove useful because of the reduced sizes of the land holdings. In most of the states the ceiling on land has been fixed at 10.117 to 12.140 ha and it is not economical for the farmers to use bigger tractors.

The conventional four-wheeled tractors can be used on such lands only on co-operative basis or on custom basis. The four-wheeled small tractors have the riding arrangement, whereas the power-tillers are walking-type tractors. In Japan, these power-tillers have become extremely popular and about eight to ten lakhs of them are in use on Japanese farms. They are also commonly used in Germany, Italy, France, etc. The power-tillers are provided with a variety of attachments, which can perform almost any agricultural operation. Already, three factories in India have started producing power-tillers with Japanese collaboration and one more factory in the co-operative sector is coming up.

Apart from the types of tractors mentioned above, there are half-track tractors as well as baby track-layers. For wetland paddy-cultivation, cage wheels have been designed. They are also called skeleton wheels. They can be broad or narrow, ranging from 0.5 to 1 m in width. They help the tractor to distribute its weight in wetland cultivation and do not allow the tractor to sink. The cage wheels are also useful in puddling.

Tractor Maintenance: The tractor is a wonderful piece of invention, but as in the case of similar other machines, its maintenance is extremely important. Without maintenance, the machine will not last long, nor will it give the expected outturn. It is also a complicated machine. Much research has gone into its development and great care is taken in manufacturing it. When it is delivered to the farmer, every care is taken to see that it is in as perfect a mechanical condition as modern science and technology can make it possible. There is one important difference between a tractor and a team of animals. Once the animals are properly fed and housed, they can take care of themselves for most of the time. But a tractor is a piece of dead metal and it is necessary for the farmer to keep it in good adjustment and repair through careful preventive maintenance. There are four factors which affect the maintenance procedures of machinery:

(1) Machinery must be designed to suit the jobs and conditions they are expected to meet in the locality;
(2) the owners and operators should have a thorough knowledge of the machines, their operation and maintenance techniques;
(3) a thorough study of the instruction books and manuals, which are usually carefully prepared and are supplied with every machine that has been purchased from a reliable manufacturer or his dealer, is necessary ; and
(4) a good spare-parts dealership and delivery service are essential.

It is not possible to cover the entire subject of machine maintenance in a general chapter of this type. All that can be attempted is the discussion of a few important principles.

The fact that the engines of a conventional tractor and that of an automobile are in many ways similar has led to many prevailing misconceptions. A car-owner, who is accustomed to send his vehicle for lubrication to the workshop once a month is inclined to under estimate the importance of the lubrication in maintenance of a tractor, as he is under the impression that a tractor being a slower-speed vehicle than his car, should not need maintenance more than his car does. It must be realized that although the travelling speed is low, the static and dynamic loads on the engine transmission and other parts of a tractor are much heavier than those in the case of a car. Also, a tractor has to work for as long as eight to ten hours daily for days and pull heavy loads. This work is equivalent to driving a fully loaded automobile at 80 km an hour for ten hours a day for days together. Hence, what would be the monthly maintenance of an automobile would become a daily chore for the tractor. It is the ignorance of this factor more than that of any other that is responsible for many of the troubles that tractor-owners have experienced in the past.

In order to get the best out of a machine, principles of preventive maintenance may be defined as the technique of maintaining a tractor at the peak of its operational efficiency throughout its working life, keeping the wear and tear to the minimum all the time, i.e. during both operation and storage. The factor responsible for mechanical deterioration is wear and tear of the various components of the machine. Wear and tear can be classified as follows:

(1) Wear due to mechanical abrasion ;

(2) corrosion due to contact and reaction with moisture, air and products of combustion of fuel, etc. and

(3) the deterioration of certain materials susceptible to ageing, fatigue, etc.

Each part of the machine may be subject to one or all of the above factors to varying degrees. Preventive maintenance aims at minimizing and counteracting the effect of the various factors responsible for the deterioration of the machine. Conditions of operation also influence the frequency and intensity of maintenance schedule. Maintenance can be considered under the following three heads:

(1) Lubrication maintenance.

(2) Operational adjustments to keep the machine in proper operational tune for peak efficiency.

(3) Maintenance during storage and off-periods.

Maintenance of Lubricant: The effect of oil in reducing friction is well known. The present-day tractor is a complicated machine and has to be properly lubricated with the right lubricant at proper intervals. Mainly, the parts requiring attention are the engine unit, the engine accessories, the clutch, the transmission, the final drive and the chasis. Every tractor is supplied with a lubrication chart, stating the number and the location of the points requiring lubrication, giving further the type of lubricant and the frequency at which attention is required. Detailed instructions are given regarding how each job is to be attended to.

It is important to see that meticulous attention is given to lubrication. A vast majority of the cases of machine breakdowns are traceable to faulty lubrication. The oils must be kept clean and changed at regular intervals. Several notable advances in tractor design have been made since the Second World War and many of the present-day designes are more compact, lighter and considerably more efficient than the older ones. Along with the improvements in the mechanical design, the lubricants and fuel technology have also advanced in no small measure, and they have contributed to the improvements in the performance of machines noticeable nowadays. Two decades ago it was a common practice to use a mineral oil straight as engine lubricant. But now, almost all the brands marketed by leading oil companies are blended with special additives to counteract certain undesirable and injurious chemical and physical changes that take place in the oil under the conditions to which it is subjected while the machine is being used.

Some substances are added to it to dissolve the carbonaceous deposits that form on the essential parts of the engine. These deposits are removed when the engine-oil is drained. This is why the improved oils turn faster than the uninhibited oils. This darkening of the oil is an indication that it contains an active detergent which is actually dissolving harmful deposits.

It is incorrect, therefore, to conclude (as is ordinarily done) that the oil has gone bad prematurely. In fact, the colour of an oil is no criterion of its quality. The special protective substance used in an inhibited oil gets used up while the machine is operating. After sometime, the oil also becomes highly contaminated. It is, therefore, very important to change it at regular intervals. The oil specifications found in some manuals, formed the basis of selection some years ago. But these are now practically valueless. The best guarantee to a consumer regarding the quality of an oil nowadays is to buy it from an authorized and reliable dealer, preferably in factory-sealed containers.

Reclaimed Oils and its Use: Some persons—even those with technical background—advocate the use of reclaimed oils. Several firms manufacture the requirement needed for the reclamation of oils. That equipment is no doubt highly efficient, but it is however, suitable only in the case of straight mineral oils, as such an equipment can remove only the solid contaminants that are held in suspension physically, but cannot remove dissolved substances. For this reason, such equipment is unsuitable for oils containing chemical additives. It is also not possible through simple chemical tests to discover what proportion of the additive substances are still left unused in the oil. Therefore, no system of oil reclamation devised so far can deal with the new types of oils. The best thing for an average user of oils to do is to discard the spent oils drained from his machine. Some dishonest dealers offer to buy back these oils at attractive prices. But since the same oils may be used for adulteration, it is dangerous to be attracted by such a temptation.

The usual grade of oil generally used in engines is SAE 30, which refers only to the viscosity or the body of the oil and gives no indication of the quality of or additives present in it. Some air-cooled engines may use an oil as thick or heavy as SAE 50.

Lubricant Transmission: The present-day tractor transmission (gear-box) and the final drive system operate at much higher loads than formerly. The working temperature of the lubricant also can be very high and under heavily loaded operating conditions, the working temperatures may rise to 60°C or more. To withstand such severe conditions, the oils for use in such transmissions also have to be specially blended. Just any thick-looking oil will not do. These oils are refined and blended with the same meticulous care as the engine-oils and some special extreme-pressure additives are also added to give the oil some special qualities. Therefore, it is best to use the recommended grades and brands purchased from reliable dealers.

The usual grade of oil specified for tractor transmissions and final drives are SAE 90 and SAE 140. Here again, the SAE specification number is no guide to the quality of the oil or to the additives present in it, since it is purely based on the physical property of viscosity. Transmission oils can also be contaminated and can deteriorate while being used and have to be changed once or twice a year. While working the machine, the oil level in the various components should be maintained at the correct levels by regular inspection and topping up. Overfilling should be avoided, as it is likely to lead to the failure of shaft seals and to other complications due to overheating and excessive frictional resistance.

Important Grease Points: Tractors are provided with several grease nipples to lubricate different parts. Greases are semi-solid lubricants of various consistencies designated as 00,0,1, 2, etc., higher numbers denoting progressively less viscous or heavier constituents. The main constituents in a grease are a metallic soap and a carefully chosen mineral oil. The main types of greases are a sodium-base grease, a calcium-base grease and mixed-base greases. The first type of grease contains soda-base soap, similar to household soap and it is water-soluble, but has a high melting-point. It is suitable for lubricating anti-friction bearings and is known as the bearing grease. The second type contains a calcium-base soap and is insoluble, but has a lower melting point. The so-called cup greases or water-pump greases fall into this category. The mixed-base greases have properties intermediate between the two and are called general-purpose greases. Greases should be carefully selected and extra care should be taken in its storage, as it easily attracts dust.

It is best to buy this lubricant in a size of packing that entails the minimum storage period under unsealed conditions. It is a very unwise policy to save money by going in for cheap brands of lubricants. The cost of the lubricant is insignificant in comparison with the cost of fuel and other expenses incidental to running a tractor. The contamination of lubricants with dust and dirt is responsible for highly accelerated rates of wear and tear. A well-maintained tractor may run for over 10,000 hours before it becomes necessary to undertake a major reconditioning job. A badly neglected tractor may be completely worn out after a few hunderd hours.

Efficient and safe operation of tractor. By proper operation, it is possible to attain economy in fuel consumption to the extent of at least 30 per cent or more. Proper attention to the routine adjustment of fan belt, valves, fuel and electrical systems will pay for the trouble taken and the expense incurred many times by increased efficiency and economy.

The majority of the tractors purchased nowadays are fitted with pneumatic tyres. These are perhaps the most expensive parts of a tractor. The most important factor affecting the life of a tyre is the maintenance of the correct inflation pressure, as stated in the instruction manual. Tyres should also he inspected periodically and any minor cuts should be promptly repaired. Nothing ruins a tyre so quickly and completely as running it in a deflated condition or even in an under inflated conditions. Therefore, the tyre pressures should be checked and maintained every day. A modern tractor is a very versatile piece of equipment and is provided with at least a belt pulley and power-take-off drive for operating various machines and devices. These devices can be highly dangerous if operated without proper guards and shields. Operators should not wear loose clothing. The best dress is close-fitting overall of *khaki* or dungaree cloth and a pair of heavy ankle boots.

Tractor-drawn Tools: The tractor-drawn implements can be classified into two categories. One set of implements is for tillage and the other for stationary operations. For tillage operations, the implements are ploughs, harrows, seed-drills, and combined harvesters and threshers. Under the stationary equipment are pumps, stationary threshers, winnowers, graders, etc.

In addition, the tractor can be used for transporting fertilizers, manures or produce by attaching a trailer to it.

The Plough: The tractor ploughs are of two kinds, namely, the mouldboard and disc ploughs. They cut two to five furrows at a time, depending on the number of complete bottoms or bodies composing the ploughs. In selecting a plough with the correct number of bodies in it, the maximum draw-bar pull of the tractor in the highest ploughing gear and the draught of the largest furrow normally required to be cut should be taken into consideration.

The correct speed of the tractor is important for good ploughing. The best speed for ploughing with normal types of ploughs is from 2.4 km/hour, depending upon the soil and other field conditions.

The depth of ploughing and burying trash. The best inversion is secured with mouldboard ploughs when the depth is less than two-thirds of the width of the cut of each bottom. A working depth of less than one-fourth of the bottom width would result in a poor quality of work. A 30-cm plough can work well up to depths varying from 8 to 20 cm. A digger-type plough cuts a furrow almost equal in depth to its width, but cannot deal so successfully with heavy trash. The shape of the share, particularly its vertical section, affects penetration.

A disc plough cuts a furrow whose width varies with the operating depth. The optimum depth is that which gives a width of cut equal to one-third the diameter of the disc. Unlike a mouldboard plough, the penetration is influenced also by the weight of the implement. So there is always some provision made for adding weight to secure the desired penetration in stiff soil. The shape of the discs, particularly their concavity, affects inversion and pulverization.

A mouldboard plough is more efficient than a disc plough if much of a surface vegetation has to be buried. But a disc plough is not liable to heavy damage from stones, roots and stumps.

The draft of a mouldboard plough varies from 0.140 to 0.21 kg/cm^2 of the cross-sectional area of the furrow slice in light soils to as much as 1.406 kg/cm^1 of extremely heavy and sticky clay soils. In summer in *kans* infested lands, the unit draft is as high as 13.607 kg.

For a 40-hp tractor, a three-bottomed plough, having 30-cm wide mouldboard and weighing about 453.592 kg is usually suitable. It will plough up to 20 cm deep and cover about 2 hectares a day.

The disc coulter is suitable for all types of ploughs ; it cuts cleaner and has a lighter draft than the knife coulter. The cut-away disc and the wavy-edged coulter are specially useful for working in a surface growth of creepers and tangled green-manure crops.

Disc ploughs are of two types, viz. disc ploughs and harrow or poly-disc ploughs. The disc ploughs are essentially heavy implements, having a few large discs (61 cm in diameter and above). They turn deep, 25 to 30.48 cm wide furrows and are best suited for cross-ploughing and stubble cultivation, particularly in hard soils. A trained worker can cover 1.6 to 2 hectares in an eight-hour day with a three-furrow disc plough or 2.24 hectares if the ploughing is light and shallow. For a 40-hp tractor, a plough having four 66-cm-diameter discs and weighing 16 to 861.825 kg is suitable. It will plough 25 to 28 cm deep.

The harrow ploughs are lighter, and have a large number of small-diameter discs for the same width of cut.

The field output of ploughs and harrows. The usual rule of thumb for estimating the output of an agricultural field operation is to multiply the effective working width of the implement in metres by the speed of the implement in kilometre per hour. The product gives the output in hectares in ten hours. This formula allows a reasonable idle time for turns on head lands, etc. Thus a 7.6 30-cm plough being pulled at a speed of 0.67 m/cm an hour can plough about 2 hectares (3x1.5) in ten hours. It is useful to remember that a speed of 6.7 m/sec is equal to a speed of 24 km/hr. Thus if with a watch you find that a tractor moves 4.87 km/hr it will cover 2 km/hr. The average pace of an adult is 76 to 83 cm. It is a good thing to know the length of one's average pace by pacing a known distance. With some practice one can measure distance by pacing with sufficient accuracy for all practical purposes. The number of 33-inch paced in 20 seconds divided by 10 gives the speed in miles per hour.

Plough Care: The daily checking of plough adjustments, the tightening of bolts and nuts on the frame and bottoms and the regular attention to lubrication and replacement of worn-out parts prolong the life of the ploughs and lessen the cost of ploughing. To ensure efficient and economic working, the hitching of the ploughs should be in the line of the draught as nearly as possible. Furthermore, the hitch should be adjusted laterally, so that the tractor tyre does not rub against the furrow-wall unduly. While sharpening the shares, their original shape should be maintained. Cast-iron shares can be re-sharpened with a power grinder. Steel shares are sharpend by hammering them after reheating. Badly worn-out, dull shares increase the draught by 10 to 20 per cent and should be replaced with new ones.

For good ploughing, the following points should always be kept in view:

1. Furrows should be straight and compact, of uniform depth and width, and with even-finished ends.

2. The furrow wall and bottom should be clean-cut and the top of the inverted furrow slices should be level to give an even seedbed.
3. The depth of ploughing should suit the soil and the crop.
4. All vegetation, manure, etc. should be properly covered.

Farmers and Harrows: In general, cultivators are used for heavy work and for breaking large clods and the harrows are used for lighter operations.

Typs of Cultivators: The former have fine pulverizing effect, whereas the latter are good for deep cultivation in light soils. The average output of a 2.4 m-wide cultivator with 11 tines is about 0.6 hectares per hour. Heavy cultivators fitted with wide shares, going 5 to 8 cm deep even in hard soils, can be used for stubble clearance. Harrows are equipped with discs or spring tines or spikes. Disc harrows are used mainly for tillage in heavy or hard soils, but can also be used for a variety of other operations, such as the covering of seed and the destruction of weeds.

The spring-tined harrows are lighter implements, weighing about 11.339 kg per tine. They can work up to a depth of 13 cm and cover about one-third of a hectare per hour, with furrows spaced 30 cm apart.

The spike-toothed harrows are mostly used for light work and can be operated with small-wheeled tractors. The medium and heavy types can also be used for preparing seedbed on uneven and rough land, for covering seed and harrowing cereals.

Among rotary cultivators, small types carried on two wheels and operated by 5- to 8-hp engines are likely to prove useful for small holdings and market gardens. They produce the combined effect of ploughing and cultivating.

Heavier rotary hoes work up to 20 cm deep and need about 6 hp per 0.304 m width on average soils. Some efficient types of small tillers suitable for rice cultivation, have appeared in foreign markets. These are being tried in this country at present.

The Roller: On light land, a set of rollers, weighing up to 610 kg and 3 to 3.6 m wide, is required to provide full load for a small tractor. The roller needed for heavy land should weigh 864 kg more. The Cambridge roller with fluted surface is very effective as a clod-crushing unit.

Seed-drills and Planters: The seed-drills can sow all kinds of small seeds, e.g. wheat, oats, barley, *jowar* and paddy, and can be adjusted for variations in the size of seed, depth of seeding, seed-rate and the distance between the rows. Drawn by wheel-type or light crawler tractors, they can sow up to about 12 ha in a day, depending on soil conditions.

Planters are employed to sow seeds in rows far enough to permit mechanical intercultivation. Depending on the crop they are designed to sow, they are called corn, cotton, peas, bean, potato, and vegetable-planters.

The trailers are of two types, three to five-tonne fixed trailers or three- to five-tonne self-tilting trailers. Depending upon the time the trailers are to be emptied, the type of trailer should be selected.

How to Maintain Machinery ?: The Indian farmers are said to be not mechanically minded. It is necessary that this quality should be imbibed in them now. Unless this is done, they may not be in a position to work the implements and machines efficiently. One of the best ways of doing this is to get rid of the idea that for every different job a different village artisan is to be sent even for carrying out minor repairs and adjustments. The old sociological systems existing in the villages and based on different castes or professions must now be changed. For major operations, no doubt, specialized persons will have to be called, but for minor repairs and adjustments, a farmer must learn to do himself. By using a set of hand-tools, he can do the jobs requiring wood-cutting, some metal work, or sheet metal work, or even work in respect of leather or planting. Thus he need not sent for either a carpenter, or a blacksmith, or a tinsmith, or a leather-worker for minor jobs.

The following sample tools are recommended for every farmer using agricultural machinery:

A wood-saw, a small drill, a hammer, three sizes of chisels, a wooden plank, some rough files, an auger, pinches, a steel file, a hacksaw, a pipe wrench, an adjustable wrench, a one-foot piece of steel girder to serve as an anvil.

He should also maintain some spare screws, nuts, bolts, washers, etc. By keeping and learning to use such a set of tools and maintaining commonly required spare parts, a farmer would be able to carry out simple repairs, such as cleaning and oiling of chaff-cutters, cane-crushers, Persian wheels, sharpening ploughs shares and cultivator shares, the tightening of nuts, bolts, etc. The knives of the chaff-cutters should be sharpened by grinding or with an ordinary file. In the case of plough shares if they are to be sharpened by heating, they should be taken to a blacksmith. The ploughshares should be of high carbon steel. A well-sharpened ploughshare reduces the plough draft by as much as 20 per cent.

After the implements are used, they should be cleaned and oiled so as to retain the polish of the steel and also to avoid their rusting. They should be always kept in a shed.

The modern farmers have also to learn how to use improved implements. In the case of indigenous implements, the construction is so simple that it is not necessary to undergo any training, but in the case of modern implements, the farmers have got to learn their use. This training may range from a single day to about two weeks training in an organized way, depending upon the type and complexity of the machine. The rule to be remembered is that with every machine, the manufacturers supply a leaflet or a booklet containing instructions for operating and maintaining it properly. It should be thoroughly studied by the farmer and should be carefully preserved for future use or reference.

The instructions in the booklet or manual should be followed in every detail. In these instructions, usually the type of lubricants to be used, their number, and the duration after which they should be used are also given. Therefore they are very important. All the bearings and other working parts should be oiled well to prevent their rusting, to secure efficient working and to prolong their life. Excessive oiling collects dirt which damages the machines. Only clean oil and grease should be applied and the oil holes should be covered. Agricultural machines are used seasonally. Therefore the farmers have plenty of time to repair or oil them. They should be stored in a roofed structure before they are required to be used in the next season.

Production of Agricultural Tools and Machines: Since Independence, quite a large number of firms in India have started manufacturing implements. There are about 150 organized industries in the country and they mass-manufacture agricultural implements.

The figure of 150 is not large and the village blacksmiths and carpenters still play a very important role in manufacturing simple agricultural implements. They will continue to do so for quite a large number of years, considering the extremely large number of farmers in India. They manufacture indigenous implements, bullock-carts, and nowadays they have also entered the field of repairing improved implements. If proper training is given to these blacksmiths and carpenters, they would prove very useful and effective in maintaining improved implements of the farmers. In order that they may shoulder their new responsibilities efficiently, the Government of India have established about 40 agricultural workshop wings attached to the extension-training centres for training the sons of blacksmiths and carpenters in repairing and maintaining improved agricultural implements. The trained artisans are also given loans for purchasing improved tools which they can use in the villages.

In the organized industries, most of the manufacture is limited to light mouldboard ploughs, turn-rest ploughs, ridgers, cultivators, hoes, wheel-hoes, olpad threshers, harrow *patelas*, diesel engines and centrifugal pumps, winnowers, sprayers, dusters, and sugarcane-crushers, tractors, power-tillers, etc. The implements are manufactured by using cast-iron, mild steel, high-carbon steel, sheet metal and wood. Most of the wood used is that of *babool* or *kikar (Acacia nilotica)* and *shisham (Dalbergia sissoo)*. High-carbon steel is used for manufacturing shares of ploughs, shovels, chaff-cutter blades, pruning-knives, etc. Cast-iron is largely used for manufacturing the frameworks of sugarcane-crushers, chaff-cutters, shelters, bodies of turn-rest ploughs and other heavy parts. The plant-protection equipment, such as sprayers, dusters are made of copper, brass or brass sheets.

Considering the demand for improved implements and machinery, some of the state governments have established their own factories in the public sector. These factories are located in Haryana, Tamil Nadu, Uttar Pradesh, Maharashtra and Rajasthan. Most of them are now under Agro-Industries Corporations of the state governments. Other state governments are taking steps to establish at least one such factory in each state. Some of them may also undertake the manufacturing of power-tillers in the future. The difficulties of the manufacturers, such as the lack of particular types of materials, e.g., sheets and steel, the provision of factory sites to them in the industrial estates, the provision of sufficient electric power at a cheaper rate, need to be removed so that the manufacturing of implements becomes easy and profitable.

The Government of India have also undertaken the standardizing of improved agricultural implements, so that their main parts may be interchanged and uniformity may be maintained. Standardization also facilitates the inspection and quality-marketing. The Indian Standards Institution has already published about 50 standards for improved implements and tools and others are under preparation. The Board for Agricultural Machinery and Implements recently established by the Government of India has recommended a number of steps, so that improved implements can be designed, tested, manufactured and demonstrated to the Indian farmers as early as possible. Action is being taken on these recommendations.

Machine and Tool Use Practautions: Every precaution should be taken in the use of machinery and implements, particularly of the power-operated types, and those having sharp working parts. In the case of chaff-cutters, the wheels should be locked when not in use to avoid' any accidents. In the case of plant-protection equipment, the equipment should be thoroughly cleaned and dried after use and the poisonous insecticides should be stored safely away from grains and other articles of food. In the case of tractors, double riding should not be allowed. No operator should ride in between the tractor and the implement.

The tractor operators should also be trained to use implements, such as ploughs, and whenever any obstruction occurs, the automatic release should be worked. Otherwise, there is a danger, that the tractor will tip. While operating the tractors and machinery, it is advisable to wear shirts and half-pants and no loose clothes, such as a *dhoti,* nor any flowing robes should be used. The fast-moving gears, transmission units, chains and spikes should be covered with guards. In the case of Olpad threshers, the discs should also be covered with expanded wire-netting. The tractor fuel, such as petrol, kerosene or diesel, should be stored very carefully. Like a small set of a mechanic's tools recommended for every farmer, a small first-aid box should be maintained. The trailers and tractors should be registered with the transport authorities and should be maintained as per rules and regulations.

It is a fallacy that the Indian farmer is slow at adopting new machines or that he does not need them. There are certain difficulties in the way of mechanization of agriculture, but they are not insurmountable. High prices of the machines mainly stand in the way of their adoption. To get over this hurdle, various methods can be adopted. Some of them have already been adopted in India.

(1) Custom-hiring by the farmers after they have completed their own agricultural operations.

(2) Custom-hiring by the agro-service centres established by the unemployed technical personnel.

(3) Custom-hiring by the Agro-Industries Corporations or other semi-government bodies.

(4) Hire-purchase.

(5) Collective farming.

(6) Starting of "Machinery Pools" as in Australia, or "Group Farms" as in Uganda in East Africa.

(7) Subsidizing the machines for the use of small and marginal farmers.

(8) Making credit available on easy terms through Land Mortegage Banks, or through I.D.A. Loans.

(9) Controlling the price of agricultural machinery and implements.

Agricultural engineering is slowly becoming a complicated subject, as more and more modern and sophisticated machines are being used. This is particularly true in the case of large land-levelling projects. It is time to recognize the utility and importance of the consultancy service in agricultural engineering and to make use of such services and expert advice while planning large and important projects. For speeding up the mechanization of agriculture, there is an urgent need for teaching the basic principles of farm machinery and implements to the boys in the schools. Therefore, simple books in regional languages on this subject need to be published for use in the schools all over India. The Government of India have approved a scheme for establishing a National Institute of Agricultural Engineering at Bhopal in Madhya Pradesh. This Institute will help to spread mechanization and make it more fruitful. A 'time will come when the mechanization of agriculture will take a big leap forward and we shall be able to help not only our farmers but also farmers in the developing countries of Asia and Africa.

12

Plant Movement

Various Kinds of Responses

Besides exhibiting overall growth, plants also show some kind of reorientation in their organs, in response to certain external stimuli such as light, gravity and water. They either bend towards or move away from the stimulus by evoking either growth or some kind of reversible change in some of their parts. This reorientation is generally termed as plant movement and is of two types: (a) Growth movements and (b) Reversible movements.

In growth movements, the change in plant due to external stimulus is plastic or irreversible and it happens due to the growth in the plant part.

In irreversible movement, there is no growth and the change is elastic, generally resulting from turgor changes in some cells.

Depending upon the direction of the plant to the stimulus, there are two types of responses or movements: (a) Tropic responses and (b) Nastic responses.

In tropic response, the reaction is related to the direction of the stimulus. The change in the organization of plant may be either in the same direction (positive) or in opposite (negative) direction, or at a specific angle to it. Examples of tropic responses are phototropism, geotropism, hydrotropism and thigmotropism.

In nastic response, the response is unrelated to the direction of the stimulus. Examples are epinasty, hyponasty, nyctinasty, seismonasty and the reactions of various types of traps in carnivorous plants.

Tropic Responses

Phototropism: Tropic movements in response to light is the phototropism. The phenomenon was perhaps studied for the first time by Charles Darwin in 1880, using canary grass and oat coleoptiles. He observed that the phototropically active light was absorbed by the tip of the coleoptile, resulting in the unequal growth in the cells on two sides below the tip of the coleoptile, causing bending of the coleoptile towards the light. F.W. Went (1928) elaborated the work of Darwin and suggested the involvement of auxins in phototropism. The phenomenon has been studied extensively since then, mainly in the etiolated grass coleoptiles, particularly in those of oats, wheat and maize and in the epicotyls and hypocotyls of light grown dicotyledonous seedlings particularly of pea and sunflower.

The stems, coleoptiles or leaves bend towards the light. This is called *positive phototropism.* The cells on the shaded or darkened side elongate more rapidly than on the illuminated side and thus the organs bend towards the source of illumination.

Many roots are also reported to respond unilaterally to the direction of light. They bend away from the light and the phenomenon is called *negative phototropism.* In this case also, this is because of the unequal growth of cells on two sides of the root. But in this case, the cells on the illuminated side elongate more rapidly than on the darkened side. However, roots of many plants are neutral to light and do not show any phototropic response.

Photoreceptor: The phototropically active light has to be absorbed by some pigment in the plant tissue to evoke responses. Among various wave lengths of light, the blue light is the most effective in causing phototropic curvature. In fact two peaks of curvature are seen in the blue region; one at 440 nm and other at 480 nm. In plants, there are two types of yellow pigments which absorb blue light, P-carotene and riboflavin. Many people are of the opinion that riboflavin or a similar flavin compound is active pigment in phototropic reception. The effective photoreceptor pigment concentration need to be very small, only about 10^{-9} M. Certain mutant organisms that have less than 2% of the normal amount of carotene are still phototropic.

In a seedling, the photoreceptor pigment may be located in coleoptile tip, as in grasses or in hypocotyl as in dicotyledonous seedlings. The hypocotyl location of the pigment in sunflower seedlings was demonstrated by S. Lam and A.C. Leopold (1966). Seedlings that had one or both cotyledons covered with black paper still showed positive phototropism, regardless of the orientation of light and the cotyledon which had been covered. Covering cotyledons reduced the intensity of the response, but not its direction. However, covering the hypocotyl prevented phototropic response, even if the cotyledons were exposed. In oat coleoptile, the tip is the most sensitive region of perception. The apical 2 μ m of the oat coleoptile is said to be about 36000 times more sensitive than the region at a distance of 2 mm from the tip.

Mechanism: The bending of the shoot or root is due to differential growth on the two sides of unilaterally exposed organs. In shoot tissues, the growth on the darkened or shaded side is more than on the illuminated side. In the phototropic responsive roots, the situation is reverse, the growth rate is retarded on the shaded side.

It has been now established that the hormone auxins were involved in unequal growth and phototropic curvature. In shoots the side which has a faster growth rate, has higher amount of auxins than the one which has slower growth rate. Earlier, the unequal distribution of auxins was thought to result from a combination of three different mechanisms: (i) Photodestruction of auxins on the illuminated side. A.W. Galston (1949) had demonstrated that riboflavin catalyses the destruction of auxins by light, (ii) Increased synthesis of auxins on the darkened side, (iii) Lateral transport of auxins from illuminated to dark side.

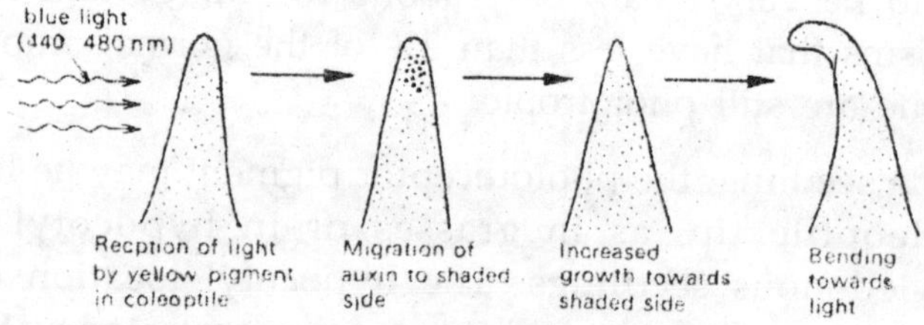

Diagrammatic representation of the events in positive phototropism in oat coleoptile.

However, now it is evident that lateral transport of auxins is the most important mechanism. In maize coleoptile, the exogenously supplied labelled (^{14}C) IAA moves away from the illuminated side to the darkened side. In pea seedlings also, blue light from one side causes a significant movement of ^{3}H-IAA but not of GA_3, another growth hormone, to the shaded side. The accumulation of IAA on the shaded side increases growth on that side, causing bending of the shoot.

How the roots respond differentially to the differential distribution of auxin is not clear.

The mechanistic events in positive phototropism of the shoots are represented diagrammatically in figure.

First Positive and the Second Positive Response: The amount or the intensity of illumination has a peculiar effect on phototropic curvature of the shoots. The positive phototropism in oat coleoptile can be induced by as low as 10^{-11} Einstein (6.6×10^{-6} cal) per cm^2 of blue light. This phototropic curvature is termed as first positive response. If the amount of illumination is increased above this intensity (10^{-11} e), the curvature is reduced and at 10^{-9} e per cm^2, it is even slightly reversed. That is, a negative phototropism can be seen, instead of positive. If the illumination increases further, again a positive phototropic curvature is initiated. This is called second positive response. In nature, phototropism must always be due to second positive response, as incident light is quite intense. Both, first and second positive responses are apparently due to the same photoreceptor pigment and to auxin migration.

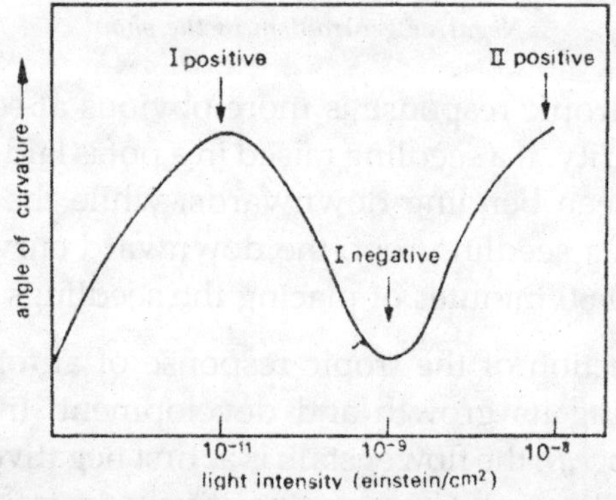

Phototropic response with increasing energies of light

Geotropism: Growth movements caused by the gravitational pull are called geotropism. Roots, specially the primary roots, are positively geotropic. Secondary roots are less geotropic than the primary roots. Tertiary roots and roots of higher orders are very little geotropic, and therefore grow almost horizontally. Stems and flowers are generally negatively geotropic. If a potted plant is kept in an inverted position for a few days, the stem can be seen bending and projecting upwards. However, the fertilized flowers of ground nut show positive geotropic response. Branches, petioles, rhizomes and stolons are usually less geotropic than the main stem and therefore they grow horizontally.

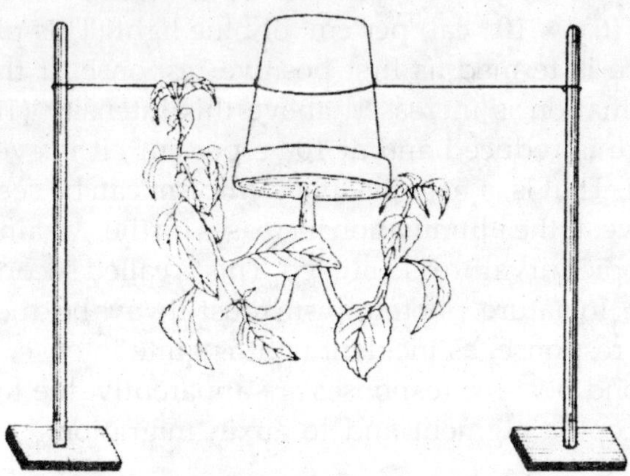

Negative geotropism in the shoot

The geotropic response is more obvious at seedling stage than at maturity. If a seedling raised in a pot is laid horizontally, the root is seen bending downwards, while the shoot bends upwards. In a seedling root, the downward curvature can be seen after 30-60 minutes of placing the seedlings horizontally.

The direction of the tropic response of an organism may change during its growth and development. In ground nut (*Arachis hypogea*), the flower stalk is at first negatively geotropic, but after fertilization it becomes positively geotropic and buries the fruit in the soil.

For a geotropic response, continuous stimulation by gravitational pull is not essential. After a period of stimulation, curvature will occur even if the test object is brought back to its original position, slowly. The minimum time during which the stimulus must be allowed to work for a subsequent geotropic curvature is known as *presentation time.* The time required for an observed response (curvature) is known as *latent time* or *reaction time.* The latent time is usually measured from the end of presentation time, so that the time of response to appear, is sum of the presentation time and reaction time. In the organisms of flowering plants, the minimum presentation time with natural gravity and at room temperature varies from 3 to 25 minutes and the reaction or latent time from 35 to 120 minutes.

Geotropic Receptor: The stimulation of gravitational pull is perceived by the tip region of the root or coleoptile, or shoot. The curvature is however, caused due to differential growth in the region of cell elongation, some distance back from the shoot or root apex. Generally the geotropic sensitivity is lost, if the root or shoot tip is removed. If the tip of a geotropically stimulated seedling is placed on a cut stump, then the curvature occurs. In dicot stems, a certain population of cells in cell elongation region below the tip, are also able to perceive the gravitational stimulus. These cells, often called starch sheath are present around the vascular bundles and are rich in amyloplasts. These amyloplasts are perhaps *statoliths,* the hypothetical particles in the receptor cells, proposed by Haberlandt and Nemec in 1900. These statoliths are believed to be the actual sensors of gravitational pull. They respond to the stimulus by moving towards the lower side, and their movement ultimately causes geotropism. The statoliths are believed to be involved in geotropic sensor in monocots also.

Mechanism: As soon as the tip receives the geotropic stimulus, some kind of message is transmitted from tip to the reactivity region *i.e.,* to the region of cell elongation. This message induces the migration of statoliths; they migrate from upper side to the lower side of the organ. This migration causes the unequal distribution of the hormones on two sides of the organ (root or shoot tip). Involvement of statoliths in unequal distribution of hormones and correlated geotropic response has been shown in many studies. Low temperature treatment of the seedling, causes the disappearance of starch grains (most likely agent for statoliths) and this leads to the loss of geotropic sensitivity as well. On return to higher temperature, geotropic sensitivity re-appears concurrently with the reappearance of the amyloplasts (starch grains). The migration of amyloplasts from upper side to the lower side of the organ is considered to be equivalent to the presentation time.

Auxins seem to be the most important hormone in geotropism. According to Cholodny-Went theory, generally the geotropism is caused by auxin accumulation on the lower side of the organ, promoting growth in the shoot and inhibiting it in the roots. The transport and unequal distribution of auxin is caused by the changing permeability of cells due to the migration of amyloplasts. Some experimental observations have been recorded regarding the role of auxins in geotropism. In maize coleoptile, lateral transport of radioactive IAA under the influence of gravity, has been demonstrated. In sunflower hypocotyls, if the endogenous IAA content is decreased by placing the hypocotyls in dark, they become insensitive to geotropism.

Other hormones also seem to be involved in geotropism. I.D.J. Phillips (1972) has shown that gibberellins might participate in the geotropism of stems. On the other hand, many plant physiologists believe that AbA is the principal hormone in geotropism, at least in roots. The hormone is present in the root tips. It becomes about three times as concentrated on lower side of the horizontally placed bean and maize roots, within 30 minutes. This may result from transport as well as more synthesis of AbA on the lower side.

Recent studies by several scientists have shown that there are rapid changes in the electrical properties of roots upon geo-stimulation. By using glass microelectrodes, H. Ishikawa and M.L. Evans (1990) have shown that in a horizontally placed moong bean seedling, the electrical potential of the cortical cells on the upperside of the root is -154 mv as compared to about -62 mv on the lower side. This difference in potential is in the elongation zone of the root. When the roots are in vertical position, no such difference is seen. The change in electrical potential may lead to unequal hormonal distribution and unequal growth causing bending of the roots.

Thigmotropism: The reaction of a plant part to the stimulus of touch (tactile, rubbing or mechanical) is called thigmotropism or haptotropism. If the movement is not directional, then it is called thigmonasty.

Thigmotropic curvature is quite apparent in the stems of climbers and in tendrils. The response is mostly positive. The entire stem of some vines grown in coils after coils around the supporting object, as is seen in many members of convolvulaceae. When the stems come in contact with the support, they respond positively by bending towards the object.

The tendrils are highly sensitive to touch, some curve within 30 seconds of a single touch. They are also able to distinguish surfaces, since they respond much more effectively to rough or textured than to smooth or soft surfaces. In nature, the thigmotropic response is quite important for many plants. Plants such as gourds, grapes and peas have tendrils, which coil around an object giving support to the climbing plant.

Receptor Organ: In tendrils, the thigmotropic response is perceived by the tip. The region of thigmotropic movement generally coincides with that of perception, but from a stimulated region the stimulus may be conducted over lengths varying from a few millimetre to many centimetres. The sensitivity of the tendril decreases from the tip back. In stem also, tip seems to be responding to the touch, although in many cases, the response is seen even after the small tip is removed.

Mechanism: The thigmotropic response in tendrils and stems is also due to unequal growth of the organ. Although no direct involvement of auxins has been demonstrated, the hormone might be playing some role as it affects tendril coiling. The movement of salts and electrolytes is probably involved in thigmotropic responses. In pea tendrils, rapid changes in ATP and inorganic phosphate takes place upon touch stimulus.

Perhaps, rapid changes in the permeability of membrane takes place, which besides facilitating the movement of these metabolites, also accelerate the water uptake and cell growth.

Hydrotropism: Response of plant organs to water content is hydrotropism. It can be seen in roots of some species, which show a directional growth towards the region of high water content in the soil. However stems, seedlings or roots of many species do not show any response to the soil water content. But positive hydrotropism has been observed in the growth of pollen tubes.

It is not definite that the growth of roots towards higher water content in the soil is a tropic response. Whether they really detect and react to the differences in water content, is not known. It is likely that they simply grow faster in the region of higher water content. Growth of the roots is stimulated in adequately moist soils and so is the growth of entire plant.

Nastic Responses

Epinasty and Hyponasty: Bending of a plant organ towards, due to more growth on its upperside is called epinasty. The phenomenon is quite apparent in petioles. In nature, the petioles bend downwards, so that the leaves assume a position, in which their tips point towards the ground. Unfolding of flowers and unrolling of fern fronds are also epinastic responses. Hyponasty is the bending of a plant organ upwards, due to more growth on the lower side of the organ.

Epi- and hyponasty may be induced by many stimuli. When stimulated by light, it is often called *photonasty*. Leaves of many dicotyledonous species exhibit photonasty. When grown in dark, the leaves curl upwards (hyponasty), but when transferred to light, the leaf lamina becomes almost horizontal (epinasty). Hyponasty in some species can be induced by gibberellin application. Nyctinasty, is also a photonastic movement due to day and night conditions.

The plant hormone, auxin, seems to be involved in epinastic curvature of petioles. Unequal distribution of auxin has been demonstrated by C.J. Lyon by using radioactive IAA in *Coleus.* In epinasty, the adaxial or morphological upperside of an organ grows more rapidly than the morphological lower side. In hyponasty, the opposite is true. It results from the translocation of different amounts of auxin in the upper and lower sides.

Thermonasty: Growth responses to the changes in temperature are called thermonasty. Such growth movements are seen in flowers, leaves and sometimes in smaller branches. The flowers of tulips and crocuses show repeated opening and closing movements of flowers in response to temperature changes. The response is quite sensitive and has been noted following a temperature change of only a fraction of a degree. The orientation of leaves in some species of *Rhododendron* is also controlled by the environmental temperature. At -15°C, the leaves are hanging almost vertically close to the stem. But at 0°C, they spread out and become almost horizontal. This movement at below freezing temperature may involve change in water tension.

The thermonastic movements are permanent growth movements resulting from a growth differential between the upper and lower surfaces of the petals or leaves. In tulips and crocus, the increase in temperature causes increased growth on the upper side of the perianth segment, so that the flowers open. Cooling increases growth on the lower surfaces, making the flowers close. Under natural conditions, this kind of reaction causes the flowers to open by day and to close by night.

Nyctinasty: The nyctinastic movements are rhythmic processes controlled by the time keeping mechanism of the plant, which is called the biological clock. Leaves are quite responsive to nyctinastic responses. In the leaflets and leaves of many leguminous plants, such as in bean and *Albizzia julibrissin,* such movements have been studied extensively. The leaves are spread over in the morning and fold in the evening. The movement is caused by the relative changes in cell sizes on opposite sides of the base of the leaflets in a shoot zone called the *pulvinus.*

The mechanism of nyctinastic movement is beginning to unfold now, in several recent investigations. The amount of plant hormone auxin, transfer of K^+ and movement of water, are some of the changes associated with nyctinasty. It is believed that during day time (beginning in the morning), large amount of auxins are produced, which are transferred largely to the lower side of the petiole. It causes preferential accumulation of K^+ in the high auxin area. Because of K^+ accumulation, water

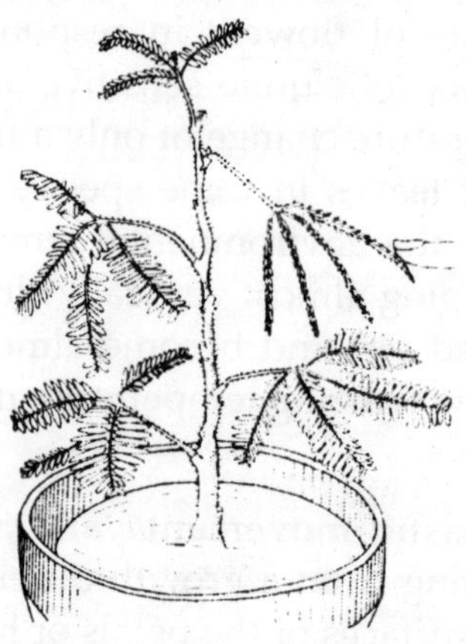

Seisminastic movement in Mimosa pudica. A mere touch has caused folding of the leaflets and drooping of branch on right side.

is translocated there, causing the rising of leaves. In the night, the auxin transport to the pulvinus is reduced and reverse reaction takes place. In experiments, auxins added to the upper or lower side of the pulvinus cause rising and falling of the leaf.

Seismonasty (Thigmonasty): Response to touch, blow or shaking is known as seismonasty. *Mimosa pudica* is the best example of seismonastic response. It responds to touch or blow by folding their leaflets and their leaves. In the leaflets, the upperside shrinks, so that the leaflets close upwards. In the petiole, the lower side shrinks, so that the whole leaf drops. The responses are very fast. The leaflet closing may start only

0.1 second after stimulation and be completed in a few seconds. These plants also respond to electrical and chemical stimuli. Only one leaflet is to be stimulated, same stimulus then moves throughout the plant, both upwards and downwards, at a rate of upto 40-50 cm per second.

Regarding the mechanism of seismonastic response, the reaction takes place in two steps: (i) transfer of stimulus and (ii) the reaction in the base of the leaflet or leaf. The transfer of the stimulus to the base of the leaf or leaflet (pulvinus) has been studied by many investigators. J.C. Bose, after extensive investigation between 1907 and 1914, proposed that the transfer of stimulus in a sensitive plant was by means of a nervous system. The proposal at that time (published in 1925) was not taken up seriously. But in early 1930s, a Dutch botanist, A.L. Houwink also demonstrated that the transfer of stimulus was in the form of some electrical impulse. Many recent studies, have also leind support to the idea of stimulus transfer through a system similar to the nervous system.

It is now established that there is the transfer of action potential in plant tissues, as it is during the nervous itnpulse in animals. The action potential apparently travels through the parenchyma cells of xylem and phloem at velocities of 2 cm per second. Barbara G. Pickard and her colleagues (1973, 1976) have found that the transfer of action potentials are widespread among plants and can be caused by several stimuli including heating, chilling, wounding or treating with salt solutions. They have also found that the long distance spread of potential depends on its movement through an unidentified chemical, called *Ricca' s factor*, after its discoverer, Ubaldo Ricca.

The stimulus after reaching the base of the leaf or leaflet, causes transport of water from the cells to the intercellular spaces. How this water is transported is unclear. However, three types of suggestions have been made:

(1) One suggestion is that the stimulus caused change in the permeability of cells rapidly, which leads to the leakage of water.

(2) According to another theory, there are large vacuoles in the pulvinus cells, which disappear after the stimulus reaches there. The vacuoles may lose water, a sort of reverse pinocytosis.

(3) Pulvinus action depends upon the hydration and dehydration of proteins. It is possible that the cell volume is changed by contraction and expansion of colloidal proteins upon the removal or addition of water. Energy for this contraction and expansion and also for the translocation of water comes from ATP, which decreases sharply during movement.

Critical experiments to establish either of these possibilities are still awaited. It is also likely that the response is combination of more than one of the above mentioned possibilities.

Other Processes

The identification is the determination of a taxon "as being identical with or similar to another and already known element". Usually the first step essential in identification of plant is the identification of the family to which it belongs. This step is accomplished by use of artificial key to families. Once the family is determined, by use of the key to genera, generic name is determined. Then by means of the key to species, specific identity of the plant is known. Identification is usually done by the utilization of taxonomic literature, floras, manuals, check lists, monographs and revisionary work on various taxa.

In identification of an unknown plant, keys are useful devices. There are various types of keys in use but the most convenient and acceptable type of key is dichotomous key. In this type of key at every stage a choice must be made between two contrasting alternatives, one of which will fit the situation and the other will not apply. Various terms used in plant morphology are explained in the glossary provided in the appendix. A brief attention is drawn here to certain features that may cause difficulty when using the key. For identification a reasonably complete specimen, with vegetative parts, flowers (if unisexual both male and female flowers) and fruits are required.

Certain deciduous plants flower when they are without leaves. In such cases leaf phyllotaxies can be determined by examining the leaf-scars. Stipules should be looked for on young shoots, as they are often deciduous. Punctate pellucid glands is another distinctive feature to look for, which can be best observed by holding a leaf against light, and so also the presence of milky or coloured sap.

The arrangement of perianth (aestivation), which is sometimes a valuable family character, should be observed in the bud, whereas placentation at full anthesis. An actinomorphic flower has sometimes only one plane of symmetry as in Brassicaceae and some Papaveraceae and a zygomorphic flower has only one plane of symmetry.

The perianth may be in one series or more series or absent. When in more than one series, it should be observed if the series resemble one another in form, colour and texture or they are differentiated in calyx and corolla. In some cases (*e.g.*, *Papaver)*, the calyx is caducous and is thrown off as the flower opens. In such cases observation in bud can give a correct idea of the situation. In some cases, as in many Apiaceae, Asteraceae and Rubiaceae, the calyx is reduced to a rim, minute teeth or hairy pappus. Such flowers are treated as having a distinct calyx and corolla. In some cases (*e.g.*, Nyctaginaceae) the calyx is petaloid structure and it should not be confused with corolla. In some cases an epicalyx is present outside the calyx (*e.g.*, some Malvaceae).

A nectar secreting disc may be present in many flowers (*e.g.*, Rutaceae) or nectar is secreted by separate glands (*e.g.*, *Geranium)* or nectar is secreted by the perianth or in sacs and spurs developed on the corolla (*e.g.*, some Ranunculaceae).

The stamens are antipetalous *i.e.*, opposite the perianth segments as in Chenopodiaceae and Amaranthaceae or the stamens of the outer whorl are opposite the petals, a condition known as obdiplostemony (*e.g.*, Caryophyllaceae).

The placentation is of vital importance in identification and it can be best observed by cutting one ovary transversely and another longitudinally by a razor blade. Attachment of ovules can be sometimes seen more easily by slitting the ovary wall with a sharp needle (*e.g.*, Asteraceae, Cyperaceae).

It is sometimes difficult to determine number of carpels in a syncarpous ovary. In such cases number of styles and stigma, the number of placentae in parietal placentation, and the number of locules in an ovary with axile placentation are indicator of basic carpel number.

It is always preferable to examine living material but floral structures can also be examined if a pressed flower is boiled in water for a couple of minutes and then dissected in a drop of water under a dissecting microscope.

With the help of the key given in the following pages, selected families of the flowering plants, can be easily identified. The key is indented type and dichotomous throughout. The main key is arranged in groups. Each pair of lead (alternative statement) is numbered, and each lead of a pair is given a distinguishing letter (a, b).

Key to the Groups

1a. Cotyledons two; leaves usually with reticulate venation; flowers usually pentamerous or tetramerous; vascular bundles of the stem usually arranged in a ring; tap root usually present. (Dicotyledons).

2a. Flowers with two distinct whorls of perianth.

3a. Corolla of distinct petals.

4a. Flowers hypogynous.

5a. A conspicuous disc is present at the base of the ovary Group 2.

5b. Disc is absent. Group 1.

4b. Flowers perigynous or epigynous. Group 3.

3b. Corolla of united petals.

6a. Ovary inferior. Group 4.

6b. Ovary superior.

7a. Carpels more than two. Group 5.

7b. Carpels two Group 6.

2b. Flowers with one whorled of perianth, usually sepaloid. Group 7.

lb. Cotyledon one; leaves usually with parallel venation ; flowers usually trimerous; vascular bundles of the stem scattered; tap root usually absent.

(Monocotyledons)

Group 8.

GROUP KEYS

DICOTYLEDONS

GROUP 1

Flowers mostly regular and bisexual; petals distinct; stamens hypogynous; ovary superior.

1a. Aquatic plants with peltate leaves.

Nymphaeaceae

1b. Terrestrial plants or if aquatic then without peltate leaves.

2a. Flowers usually trimerous.

3a. Trees or shrubs with exstipulate leaves; connective distinctive, usually enlarged; endosperm of the seed ruminate. Annonaceae

3b. Trees or shrubs with stipulate leaves; connective not enlarged; endosperm of the seed watery fleshy.

Magnoliaceae

2b. Flowers tetramerous or pentamerous, not trimerous.

4a. Gynoecium of many apocarpous carpels.

Ranunculaceae

4b. Gynoecium of few syncarpous carpels.

5a. Ovary unilocular.

6a. Placentation parietal.

7a. Flowers zygomorphic

8a. Leaves exstipulate. Fumariaceae

8b. Leaves stipulate. Violaceae

7b. Flowers actinomorphic.

9a. Stamens indefinite. Papaveraceae

9b. Stamens definite.

10a. Stamens tetradynamous. Brassicaceae

10a. Stamens not tetradynamous.

11a. Sepals and petals four each; androphore and gynophore usually present. Capparidaceae

11b. Sepals and petals five each; androphore and gynophore absent. Violaceae

6b. Placentation free central.

12a. Stamens obdiplostemonous; capsule dehiscing by valves. Caryophyllaceae

12b. Stamens not obdiplostemonous; capsule dehiscing transversely. Portulacaceae

5b. Ovary multilocular.

13a. Anthers one-celled. Malvaceae

13b. Anthers two-celled.

14a. Stamens 5-20, more or less connate. Sterculiaceae

14b. Stamens many, more or less distinct. Tiliaceae

GROUP 2

Sepals distinct or united; a conspicuous disc is present; stamens usually definite, inserted upon or at the outer or inner base of the disc; ovary superior.

1a. Leaves aromatic, glandular punctate. Rutaceae

1b. Leaves neither aromatic nor glandular punctate.

2a. Climbers, usually with tendrils. Vitaceae

2b. Trees, shrubs or herbs without tendrils.

3a. Flowers bisexual, disc usually annular adnate to the stamens.

4a. Leaves simple and stipulate; fruit a schizocarp. Geraniaceae

4b. Leaves compound and exstipulate; fruit a capsule or drupe.

5a. Stamens obdiplostemonous; may be united only at the base. Oxalidaceae

5b. Stamens not obdiplostemonous; united into a tube around the style. Meliaceae

3b. Flowers usually unisexual, disc tumid, adnate to the base of calyx or lining its tube.

6a. Fruit capsular or indehiscent, sometimes winged. Sapindaceae

6b. Fruit a one-celled one-seeded drupe. Anacardiaceae

GROUP 3

Sepals united, rarely free; often adnate to ovary; stamens perigynous, usually inserted on or beneath the outer margin of the disc; ovary often inferior.

1a. Ovary of one carpel with marginal placentation; fruit a legume. Fabaceae

1b. Ovary of two or more carpels; placentation not marginal; fruit never a legume.

2a. Herbs climbing with the help of tendrils. Cucurbitaceae

2b. Herbs, shrubs or trees without tendrils.

3a. Leaves stipulate. Rosaceae

3b. Leaves exstipulate.

4a. Placentation parietal. Cactaceae

4b. Placentation not parietal.

5a. Ovary unilocular with two to five pendulous ovules. Combretaceae

5b. Ovary usually two or more locular.

6a. Leaves with pellucid aromatic glands. Myrtaceae

6b. Leaves without pellucid aromatic glands.

7a. Fruit a many-seeded capsule. Lythraceae

7b. Fruit a schizocarp splitting into two mericarps. Apiaceae

GROUP 4

Corolla gamopetalous; stamens usually as many as corolla lobes; ovary inferior.

1a. Leaves opposite and stipulate. Rubiaceae

1b. Leaves alternate and exstipulate. Asteraceae

GROUP 5

Corolla gamopetalous; stamens as many or twice as many as the corolla lobes; carpels more than two; ovary superior. Sapotaceae

GROUP 6

Corolla gamopetalous; stamens as many as or fewer than corolla-lobes; carpels two; ovary superior.

1a. Flowers actinomorphic; stamens as many as the corolla-lobes

2a. Leaves usually opposite.

3a. Pollen granular, transferred directly from anthers; corona absent. Apocynaceae

3b. Pollen often in pollinia, transferred by means of specialized translators; corona present. Asclepiadaceae

2b. Leaves generally alternate.

4a. Ovules one or two per loculus.

5a. Sepals free; corolla lobes contorted and infolded. Convolvulaceae

5b. Sepals connate; corolla lobes imbricate. Boraginaceae

4b. Ovules numerous in each loculus. Solanaceae

1b. Flowers zygomorphic; corolla usually two-lipped; stamens fewer than corolla lobes.

6a. Ovules many in each locule.

7a Trees, shrubs or climbers with compound leaves; seeds winged. Bignoniaceae

7b. Herbs or shrubs with simple leaves, seeds wingless.

8a. Capsule opening elastically; seeds usually on hooked funiculus. Acanthaceae

8b. Capsule not elastic; seeds not on hooked funiculus.

9a. Flowers axillary; corolla ventricose, sub-two-lipped. Pedaliaceae

9b. Flowers in racemes or spikes; corolla not ventricose, usually two-lipped. Scrophulariaceae

6b. Ovules one or two in each locule.

10a. Style gynobasic; fruit usually of four one-seeded nutlets: corolla strongly bilabiate; calyx often two-lipped. Lamiaceae

10b. Style terminal; fruit a berry or drupe; corolla less strongly zygomorphic; calyx nearly actinomorphic, not two-lipped Verbenaceae

GROUP 7

Flowers usually with one whorl of perianth, usually sepaloid, or none.

1a. Stems jointed; leaves reduced to whorls of scales. Casuarinaceae

1b. Stems not jointed, leaves expanded.

2a. Ovary inferior; branch parasites. Loranthaceae

2b. Ovary superior; not branch parasites.

3a. Flowers usually bisexual or polygamous.

4a. Leaves stipulate; stipules sheathing. Polygonaceae

4b. Leaves exstipulate.

5a. Perianth scarious; stamens usually with interposed staminodes. Amaranthaceae

5b. Perianth sepaloid; staminodes not present. Chenopodiaceae

3b. Flowers usually unisexual.

6a. Ovary trilocular; fruit schizocarpic; seeds carunculate. Euphorbiaceae

6b. Ovary unilocular; fruit a syncarp of confluent pericarps and perianths; seeds not carunculate. Moraceae

MONOCOTYLEDONS

GROUP 8

1a. Trees, shrubs or scramblers with large plicate or pinnately divided leaves; flowers nearly sessile, in fleshy spikes or panicles with spathe-like bracts. Arecaceae

1b. Not as above.

2a. Aquatic herbs, carpels free. Alismaceae

2b. Terrestrial plants, carpels usually fused.

3a. Ovary inferior.

4a. Flowers in spadix. Musaceae

4b. Flowers not in spadix.

5a. Placentation parietal. Orchidaceae

5b. Placentation not parietal, usually axile.

6a. Fertile stamen one, the others often transformed into petal-like staminodes. Zingiberaceae

6b. Fertile stamens six, in two trimerous whorls.
Amaryllidaceae

3b. Ovary superior.

7a. Perianth well developed.

8a. Leaf bases sheathing, filaments bearded.
Commelinaceae

8b. Leaf bases not sheathing; filaments not bearded.
Liliaceae

7b. Perianth scarious or reduced to bristles, hairs or lodicules or none.

9a. Flowers usually unisexual, on a spadix; fruit a berry.
Araceae

9b. Flowers uni-or bisexual; in spikelets; fruit indehiscent.

10a. Leaf phyllotaxis 1/2; leaf sheaths with usually free margins; culms cylindrical; spikelets subtended by a pair of sterile bracts (glumes); each flower enclosed between a membranous bract (lemma) and adaxial bracteole; perianth reduced to lodicules.
Poaceae

10b. Leaf phyllotaxis 1/3; leaf sheaths closed; culms usually triangular; spikelets usually not subtended by sterile bracts; each flower subtended by a single membranous bract; perianth represented by bristles, hairs, scales, or none.
Cyperaceae

Bibliography

Adam, N. : *Encyclopaedia of Plant Physiology,* Springer Verlag, Berlin, 1975.

Alam, B. : *Advanced Plant Physiology,* Longman, London, 1984.

Anderson, Edgar : *Plants, Man and Life,* University of California Press, California, 1967.

Arvill, R. : *Man and Environment : Crisis and the Strategy of Choice,* Penguin, New York, 1967.

Bailey, L. H. : *Hand Book of the Multiplication of Plants,* Discovery Publishing House, New Delhi, 1999.

Beevers, L. : *Nitrogen Metabolism in Plants,* Edward Arnold, London, 1976.

Bhatia, B. M. : *Poverty, Agriculture and Economic Growth,* Vikas Publishing House, New Delhi, 1977.

Billihgs, W. D. : *Plant, Man and the Ecosystem,* MacMillan, London, 1971.

Bowling, D. J. F. : *Uptake of Ions by Plant Roots,* Chapmann and Hall, London, 1976.

Bray, C. : *Nitrogen Metabolism in Plants,* Longman, London, 1984.

Candolle, Alphonse De : *Origin of Cultivated Plants,* Hafner, New York, 1886.

Crafts, A. S. and C. E. Crisp, : *Phloem Transport in Plants,* W.H. Freeman, San Francisco, 1973.

Dadiani, N. K. : *The Survey of Indian Agriculture,* Kasturi & Sons Ltd., Chennai, 1999

Deckock, P. C. : *Mineral Nutrition of Plants : a Symposium,* University of Oxford, New York, 2002.

Dennis, D. T. and Turpin, D. H. : *Plant Physiology, Biochemistry and Molecular Biology,* Longman, London, 1990.

Douglas, J. S. : *Advanced Guide to Hydroponics,* Pelham Books, London, 1976.

Dubey, R. S. : *Agricultural : Issues and Application,* Gyan Books, New Delhi, 1987.

Dunn, Edgar S. : *Location of Agricultural Production,* University of Florida Press, New York, 1954.

Evans, L. T. : *Day Length and the Flowering of Plants,* W.A. Benjamin, California, 1975.

Farmer, J. B. : *Study of Plant Life,* Discovery Publishing House, New Delhi, 1993.

Frankel, F. R. : *India's Green Revolution : Agriculture Costs of Land Growth,* Princeton University Press, Princeton, 1971.

Ftzler M. E. : *Plant Lectins : Molecular and Biological Aspects,* Academic Press, New York, 1985.

Gangawane, L. V. : *Vistas in Mycology and Plant Pathology,* Commonwealth Publishers, New Delhi, 2000.

Gauch, H. G. : *Inorganic Plant Nutrition,* Hutchinson and Ross, London, 1972.

Green, J. R. : *An Introduction to Vegetable Physiology,* MacMillan, London, 1911.

Gregor, Howard F. : *Geography of Agriculture : Themes in Research,* Prentice Hall, New Jersey, 1970.

Gupta, G.P. : *Test Book of Plant Diseases,* Discovery Publishing House, New Delhi, 2001.

Hansen, E. : *Post Harvest Physiology of Fruits,* Praeger Publishers, New York, 1966.

Hewitt, E. J. and Cutting, C. V. : *Nitrogen Assimilation of Plants,* Academic Press, London, 1982.

Hussain, M. : *Systematic Agricultural Geography,* Rawat Publications, Jaipur, 1999.

Isser, G. P. : *Changing Face of Poverty and Globalisation,* Gyan Books, New Delhi, 2002.

James, A.: *Weather and Agriculture,* Pergamon, Oxford, 1967.

Joshi, H. L. : *Agricultural of India,* Rawat Publications, Jaipur, 1990.

Karanth, Apoorva : *Plant Genetics and Food Crop Science,* Dominant Books, New Delhi, 1999.

Kirkby, E. A. : *Nitrogen Nutrition of Plants,* University of Leeds, Leeds, 1970.

Kramer, A. L. : *Principles of Biochemistry,* Worth Publishers, New York, 1982

Kramer, P. J. : *Plant and Soil Water Relationship : A Modern Synthesis,* McGraw Hill Book Company, New York, 1969.

Kumar, Arun : *Agricultural Development : Today and Tomorrow,* Anmol Publications, New Delhi, 1995.

Kumar, T. : *History of Rice in India : Mythology, Culture and Agricultural,* Gyan Books, New Delhi, 1988.

Laxmi Narasaiah, M. : *Agricultural Production,* Discovery Publishing House, New Delhi, 1999.

Mazliak, P. : *Lipid Metabolism in Plants,* Chapmann and Hall, London, 1973.

McKee, H. S. : *Nitrogen Metabolism in Plants,* Oxford University Press, London, 1962.

McNeil, M. Darvill : *Structure and Function of the Primary Cell Walls of Plants,* McGraw, London, 1984.

McRae, S. C. and Burnham, C. P. : *Land Evaluation,* Oxford University Press, New York, 1981.

Meidner, H. and Sheriff, D. W. : *Water and Plants,* Blackie and Sons, Glasgow, 1976.

Mengel, K. and Kirkby, E.A. : *Principles of Plant Nutrition,* Praeger Publishers, New York, 1974.

Mishra, S. R. : *Bacterial Plant Diseases,* Discovery Publishing House, New Delhi, 2003.

Misra, B., Kar, G. C. and Misra, S. N. : *Agro-Industries and Economic Development,* Deep & Deep Publications, New Delhi, 2002.

Mitter, V. : *Growth of Urban Informal Sector,* Deep & Deep Publications, New Delhi, 1988.

Mukherji, Shekhar : *Land-Use Change, Environment and Sustainable Development,* Banaras Hindu University, Varanasi, 2000.

Nomura, M. : *The Control of Ribosome Synthesis,* Edward Arnold, London, 1986.

Novikoff, A. B. : *The Endoplasmic Reticulum : a Cytochemist's View,* Oxford, New York, 1976.

Numa, S. : *Fatty Acid Metabolism and its Regulation,* Janis, London, 1984.

Nutman, P. S. : *Symbiotic Nitrogen Fixation in Plants,* Cambridge University Press, Cambridge, 1976.

Ogawa, Y. and R. W. King : *Plant and Cell Physiology,* Springer-Verlag, Berlin, 1990.

Opik, H. : *The Respiration of Higher Plants,* Edward Arnold Publishers, London, 1980.

Palladin, V. I. : *Plant Physiology,* Arihant Publishers, Jaipur, 1988.

Pannikkar, K. M. : *India and the Indian Ocean : An Essay on the Influence of Sea Power on Indian History,* George Allen and Unwin, London, 1951.

Ramshaw, J. A. M. : *Structure of Plant Proteins,* Oxford University Press, New York, 1982.

Robert, W. : *Middle America : Its Land and Peoples*, Prentice Hall, Englewood Cliff, 1989.

Roberts, J. A. and Tucker, G. A. : *Ethylene and Plant Development*, Butterworth, London, 2001.

Rubenstein, J. M., and Bacon, R. S. : *The Cultural Landscape : An Introduction to Human Geography*, Prentice Hall, New Delhi, 1990.

Salisbury, F. B. and C. W. Ross, : *Plant Physiology*, Woodworth Publishing Co., California, 1967.

Samanta, R. K. : *Agricultural Estension in Changing World Perspective*, Uppal Publishing House, New Delhi, 1995.

Shandhya, T. K. : *Agricultural Credit and Nabard*, Deep & Deep Publications, New Delhi, 2003.

Sharma, D. N. : *Minerals of India*, National Book Trust, New Delhi, 1994.

Sharma, S. S. : *Sustainable Agriculture : Poverty and Food Security*, Rawat Publications, Jaipur, 2002.

Sharma, V. K. : *Trees and Protection of Environment*, Deep & Deep Publications, New Delhi, 2004.

Siegelman, H. W. and G. Hind : *Photosynthetic Carbon Assimilation*, Plenum, New York, 1978.

Singh, D. K. : *Technological Change and Agricultural Development*, Deep & Deep Publications, New Delhi, 1993.

Singh, Harjinder : *Agricultural Problems in India*, Gyan Books, New Delhi, 1988.

Singh, L. : *Green Revolution and Cropping Pattern*, Deep & Deep Publications, New Delhi, 1993.

Singh, Surendra : *Agricultural Development in India : A Regional Analysis*, Kaushal Publications, Shillong, 1994.

Slatyer, R. O. : *Plant Water Relationship*, Academic Press, New York, 1967.

Somani, K. K. : *Indian Economics and Social Traditions*, Gyan Books, New Delhi, 2002.

Srivastava O. S. : *Agricultural Economics*, Rawat Publications, Jaipur, New Delhi, 1996.

Steward, F. C. : *Plant Physiology, A Treatise*, Academic Press, New York, 1963.

Street, H. E. and H. Opik : *The Physiology of Flowering Plants*, Edward Arnold, London, 1970.

Subrahmanya, S. : *Studies in Agricultural Development*, Deep & Deep Publications, New Delhi, 1996.

Sudhir, M. : *Applied Biotechnology and Plant Genetics*, Dominant Books, New Delhi, 2000.

Thomas, C. : *A History of the Land and Agriculture*, Prentice Hall, Englewood Cliff, 1978.

Valcarcel, M. : *Weather and Agriculture*, Oxford University Press, London, 1997.

Varner, J. E., Flint D. and R. Mitra, : *Characterization of Protein Metabolism in Cereal Grains*, Academic Press, New York, 1976.

Venkateshwarlu : *Developing Agricultural Technology*, Rawat Publications, Jaipur, 1998.

Verma, M. : *The Survey of Indian Agriculture*, Kasturi & Sons, New Delhi, 1999.

Verma, P. C. : *Surplus Manpower in Agriculture and Employment Policy*, Deep & Deep Publications, New Delhi, 1991.

Vyas, V. S. : *Policies for Agricultural Development*, Rawat Publications, Jaipur, 1997.

Wilkins, M. B. : *The Physiology of Plant Growth and Development*, McGraw Hill book, London, 1969.

Zeevart, J. A. D. : *Physiology of Flower Formation*, McGraw Hill, London, 1976.

Zimmermann, M. H. : *Encyclopeadia of Plant Physiology*, Academic Press, New Work, 1982.

INDEX

D

E

F

G

N

O

P

Q

❑❑❑